9·21·93

OPERATIONAL AMPLIFIERS: INTEGRATED AND HYBRID CIRCUITS

OPERATIONAL AMPLIFIERS: INTEGRATED AND HYBRID CIRCUITS

GEORGE B. RUTKOWSKI, P.E.
Computek

A Wiley-Interscience Publication

JOHN WILEY & SONS, INC.

New York • Chichester • Brisbane • Toronto • Singapore

Library of Congress Cataloging in Publication Data:

Rutkowski, George B.
    Operational amplifiers: integrated and hybrid circuits / by George
B. Rutkowski.
        p.   cm.
    "A Wiley-Interscience publication."
    Includes index.
    ISBN 0-471-57718-9 (cloth)
    1. Operational amplifiers.   2. Differential amplifiers.
3. Integrated circuits.   I. Title.
TK7871.58.06R88   1993
621.39′5—dc20   92-20592

CONTENTS

PREFACE

For scientific and engineering applications, analog computers preceded digital computers. Analog computers were able to perform complex calculations with relatively few parts, an important advantage in the days when the parts were much larger and far less reliable than they are today. The basic building blocks of these analog computers were operational amplifiers (Op Amps). The early Op Amps were large, bulky vacuum tube circuits. As transistors replaced vacuum tubes, Op Amps became smaller, more reliable, and less expensive. These features, combined with the flexibility and ease of use of Op Amps, catapulted them into many industrial and military applications. The demands of the space age stimulated development and growth of integrated circuits (ICs) that offer more size reductions and much improved reliability. A large variety of IC Op Amps are now available. They are inexpensive and offer circuit designers virtually unlimited applications.

While digital ICs change convulsively with time, IC Op Amps are relatively stable. As a near-perfect electronic device for the niche it fills, the IC Op Amp has reached an evolutionary plateau. Metaphorically, we can say that there is an island of stability in the turbulent sea of electronics, the isle of the IC Op amp. Since old and new applications abound with Op Amps, you can rely on the knowledge that you gain here to be useful for many years to come.

For more than two decades, IC Op Amps have been mainly small-signal devices. This means that the signal power levels out of IC Op Amps are quite small; typically hundreds of milliwatts at best. In applications where signal power levels need to be larger, Op Amps are followed by power amplification. This typically includes one or more power amplifier stages that consist of discrete parts. Or, as is now the trend, the power stages and the preceding Op Amp are all placed into one package. This mix of ICs and discrete parts is called a hybrid circuit. A number of manufacturers are producing quality hybrid power Op Amps which are also discussed in this text.

In summary, this text was developed to:

1. provide an understandable and sufficiently detailed explanation of the IC and hybrid Op Amps for practicing or student engineers, technologists, and technicians;
2. serve as a text with abundant examples and end-of-chapter problems; and
3. be a reference by including a collection of Op Amp circuit applications with guidelines to selecting their component values and by providing manufacturers' data sheets with tables of Op Amp types and their comparative characteristics.

☐ Chapter 1 of this book contains a review of bipolar junction transistors (BJTs) and of unipolar transistors (FETs), their use in differential amplifiers, and the fundamentals of the circuitry inside an IC Op Amp. This chapter can be omitted if a functional block approach to Op Amps is preferred.

☐ Op Amps have a distinctive language of their own. Chapter 2 provides definitions and explanations of important Op Amp characteristics and compares their real practical values to hypothetical ideal ones.

☐ Chapters 3–7 contain detailed discussions of the significance of IC Op Amp parameters in common practical circuits. Manufacturers' data (spec) sheets are provided and frequently referenced in the way engineers and technologists are required to do in practice.

☐ Chapters 8–11 show and explain a variety of practical circuit applications with methods of selecting circuit component values emphasized.

☐ Chapter 12 discusses power amplification and the principles of power Op Amps. Power Op Amps are expensive and work at relatively high energy levels. Therefore, their proper use, including precautions and protections, is emphasized.

☐ When the source of an equation is relatively simple, it is explained in the text. Where the derivations are more complex or lengthy, they are given in the Appendices. These Appendices also include the manufacturers' Op Amp specifications (specs). Many application circuits are included also. BASIC programs that can plot gain versus frequency characteristics of active filters are listed. They can be used in conjunction with appropriate examples and end-of-chapter problems.

☐ A very thorough glossary of terms related to Op Amps and their applications is provided. This is useful, not only when learning the principles of Op Amps while progressing through this text, but also as a handy reference for years after.

GEORGE B. RUTKOWSKI

Cleveland, Ohio
January 1993

OPERATIONAL AMPLIFIERS:
INTEGRATED AND HYBRID CIRCUITS

1

DIFFERENTIAL AND OPERATIONAL AMPLIFIERS

The differential amplifier, as its name implies, amplifies the difference between two input voltages. These input voltages are applied to two separate input terminals with respect to ground or a common. Their difference is called the *differential input voltage* V_{id}. The differential amplifier has two output terminals, and output signals can be taken from either output with respect to ground or across the two terminals themselves. The signals across the output terminals are usually amplified versions of the differential input voltages. Differential amplifiers, and variations of them, are found in many applications: measuring instruments, transducer amplifiers, industrial controls, signal generators, digital-to-analog (D/A) converters, and analog-to-digital (A/D) converters, to name just a few.

In this chapter we will first consider the construction and operating fundamentals of junction transistor and field-effect transistor differential amplifiers. Then we will see how differential amplifiers can be cascaded, as they are in typical *integrated circuits* (ICs), to obtain very high gain. Finally, we will see how level-shifting circuitry is added to modify the two output terminals of a differential-type circuit to a single output from which a signal can be taken with respect to ground or a common. Level shifting changes a differential amplifier into an *operational amplifier* (Op Amp).

1.1 TRANSISTOR REVIEW

Since bipolar junction transistors (BJTs) and field-effect transistors (FETs) are frequently used in discrete and IC differential and operational amplifiers, a review of these devices is helpful in developing an understanding of the inner workings of linear ICs.

The *junction transistor* is a current-operated device with three terminals: an emitter E, a collector C, and a base B. In most applications, the relatively small base current I_B is controlled or varied by a signal source. The varying base current I_B in turn controls or varies a much larger collector current I_C and emitter current I_E.

Properly biased transistors are shown in Fig. 1-1. The term *bias* refers to the use of proper dc voltages on the transistor that are necessary to make it work as an amplifier. As shown in Fig. 1-1, the collector C is biased positively

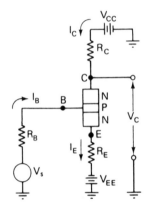

(a) The NPN transistor contains a P-type semiconductor between two N-type materials.

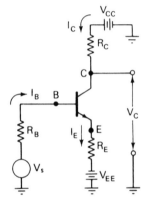

(b) Arrow on emitter points out on NPN transistor symbol.

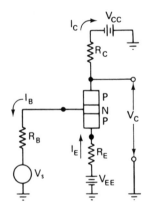

(c) The PNP transistor contains N-type semiconductor between two P-type materials.

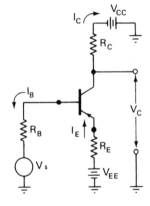

(d) Arrow on the emitter points in on PNP transistor symbol.

Figure 1-1   Properly biased transistor circuits.

with respect to the emitter E on NPN transistors, whereas the collector C is normally negative with respect to the emitter E on PNP transistors. In either case, a NPN or PNP circuit, the base-emitter junction of N- and P-type semiconductors is forward biased. This means that V_{EE} is applied with a polarity that will cause current I_B across the base-emitter junction. The values of V_{EE}, V_S, R_E, and R_B determine the value of I_B.

As base current flows, it causes a collector current flow I_C that is about β^* or h_{FE} times larger. That is,

$$I_C \cong \beta I_B \qquad\qquad (1\text{-}1)$$

or

$$I_C \cong h_{FE} I_B, \qquad\qquad (1\text{-}2)$$

where β or h_{FE} is specified by the transistor manufacturer. In modern transistors, h_{FE} typically ranges from about 60 to 200. Thus, the base current I_B is typically smaller than the collector current I_C by a factor between 60 and 200.

Note in Fig. 1-1 that the sum of the currents I_B and I_C is equal to I_E. That is,

$$I_E = I_C + I_B. \qquad\qquad (1\text{-}3)$$

Since I_C is typically much larger than I_B, the base current I_B is often assumed negligible compared to either I_C or I_E. Therefore, Eq. (1-3) can be simplified to

$$I_E \cong I_C. \qquad\qquad (1\text{-}4)$$

Variations of the input voltage V_s on any of the circuits of Fig. 1-1 will cause variations of I_B, which in turn varies I_C flowing through R_C. Thus the voltage across R_C and the output voltage V_C will vary too. Usually, the output variations of V_C are much larger than the input variations of V_s, which means that these circuits are capable of voltage gain—also called voltage amplification A_v.

The *field-effect transistor*, FET, unlike the junction transistor, draws negligible current from the signal source V_s and therefore is referred to as a voltage-operated device. It has three terminals: a source S, a drain D, and a

---

*A transistor's beta (β) is about equal to its h_{FE}. In the symbol h_{FE}, the h means *hybrid* parameter, F means *forward* current transfer ratio, and E means that the *emitter* is common. The capital letters in subscript in h_{FE} specify that this is a dc parameter or dc beta.

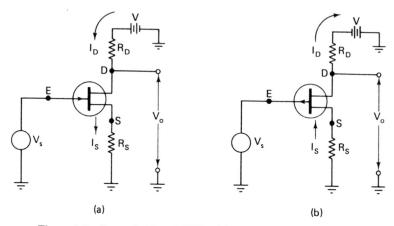

Figure 1-2   Properly biased FETs: (a) N channel, (b) P channel.

gate G. Properly biased FETs are shown in Fig. 1-2. Basically, variations of a voltage across the gate G and the source S cause drain current I_D and source current I_S variations. Thus, a varying input signal V_s causes a varying gate-to-source voltage V_{GS} which in turn varies I_D, the drop across R_D, and the output voltage V_o. The ratio of the change in drain current I_D to the change in gate-to-source voltage V_{GS} is the FET's transadmittance* g_m; that is,

$$g_m \cong \frac{\Delta I_D}{\Delta V_{GS}}. \qquad (1\text{-}5)$$

Because of the very high gate input resistance of the FET, negligible gate current flows, and therefore the drain and source currents are essentially equal; that is, $I_D = I_S$. FET amplifiers are capable of voltage gain, but not as high as junction transistors. FETs are used where extremely high input resistance is important, as in some applications of differential and operational amplifiers. As we will see, some types of Op Amps on ICs have FETs in their first stage.

FETs are made in two general types: (1) junction field effect transistors, JFETs, whose input resistances are on the order of 10^6 to 10^9 ohms, and whose symbols appear in Fig. 1-2, and (2) metal-oxide silicon field effect transistors, MOSFETs, which are also known as insulated gate field-effect transistors, IGFETs. MOSFETs or IGFETs have higher gate input resistance values than do JFETs, typically 10^{10} to 10^{14} ohms. Some common symbols in use for MOSFETs are shown in Fig. 1-3.

---

*The forward transadmittance, sometimes called *transconductance*, is referred to by several symbols: y_{fs}, g_{fs}, g_m, and g_{21}.

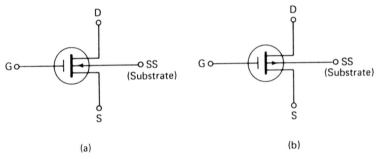

Figure 1-3   Symbols used to represent MOSFETs or IGFETs: (a) N channel, (b) P channel.

1.2   THE DIFFERENTIAL CIRCUIT

Figure 1-4a shows a simple BJT differential amplifier, and Fig. 1-4b, its typical symbol. If signal voltages are applied to input terminals 1 and 2, their difference V_{id} is amplified and appears as V_{od} across output terminals 3 and 4. Ideally, if both inputs are at the same potential with respect to ground or a common point, causing an input differential voltage $V_{id} = 0$ V, the differential output voltage $V_{od} = 0$ V too, regardless of the circuit's gain.

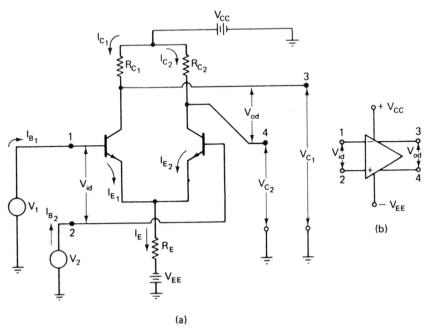

Figure 1-4   (a) Simple differential amplifier, (b) symbol for the differential amplifier.

With input signal sources V_1 and V_2 applied as shown in Fig. 1-4, each base has a dc path to ground, and base currents I_{B_1} and I_{B_2} flow. If the average voltages and internal resistances of the signal sources V_1 and V_2 are equal, then equal base currents flow. This causes equal collector currents $I_{C_1} = I_{C_2}$ and equal emitter currents I_{E_1} and I_{E_2}, assuming that the transistors have identical characteristics. Since β or h_{FE} of a typical transistor is much larger than 1, the base current is very small compared to the collector or the emitter current, and therefore the collector and emitter currents are about equal to each other [see Eqs. (1-3) and (1-4)]. In this circuit, the current I_E in resistor R_E is equal to the sum of currents I_{E_1} and I_{E_2}. This current can be approximated with the equation

$$I_E = \frac{V_{EE} - V_{BE}}{R_E + R_B/2h_{FE}}, \qquad (1\text{-}6)^*$$

where V_{BE} is the dc drop across each forward-biased base-emitter junction, which is in the range of about 0.5 to 0.7 V in silicon transistors and about 0.1–0.3 V in germanium transistors, and

R_B is the dc resistance seen looking to the left of either input terminal 1 or 2. In the circuit of Fig. 1-4, R_B is the internal resistance of each signal source.

Since the dc source voltage V_{EE} is usually much larger than the base-emitter drop V_{BE}, and since R_E is frequently much larger than $R_B/2h_{FE}$, Eq. (1-6) can often be simplified to

$$I_E \cong \frac{V_{EE}}{R_E}. \qquad (1\text{-}7)$$

The significance of Eq. (1-7) is that the value of I_E is determined mainly by the values of V_{EE} and R_E and that if V_{EE} and R_E are fixed values, the current I_E is practically constant. Ideally, I_E should be very constant for reasons we will see later. The source V_{EE} and resistor R_E, or their equivalent, are often represented with the symbol for a constant-current source shown in Fig. 1-5b.

The dc voltage to ground at each output terminal is simply the V_{CC} voltage minus the drop across the appropriate collector resistor. Thus the voltage at

---

*See Appendix A for derivation.

output 3 to ground, in Fig. 1-4, is

$$V_{C_1} = V_{CC} - R_{C_1}I_{C_1}. \qquad\text{(1-8a)}$$

Similarly, the voltage at output 4 to ground is

$$V_{C_2} = V_{CC} - R_{C_2}I_{C_2}. \qquad\text{(1-8b)}$$

Their difference is the output differential voltage

$$V_{od} = V_{C_1} - V_{C_2}, \qquad\text{(1-9)}$$

just as the input differential voltage is

$$V_{id} = V_1 - V_2. \qquad\text{(1-10)}$$

With I_E constant, as in the circuit of Fig. 1-4, the sum of the collector currents $I_{C_1} + I_{C_2}$ is also constant. Thus, if I_{C_1} is increased, I_{C_2} is forced to decrease, and vice versa. In other words, if the input voltage V_1 is made more positive, base current I_{B_1} increases, increasing I_{C_1} and the voltage drop across R_{C_1}. This in turn causes the voltage V_{C_1} at output 3 to ground to decrease. And the increase in I_{C_1} forces I_{C_2} to decrease, which decreases the drop across R_{C_2} and increases the voltage V_{C_2} at output 4. On the other hand, if V_1 is made more negative than V_2, causing a larger differential input voltage V_{id} of opposite polarity, then output 3 becomes more positive while output 4 becomes more negative.

The ratio of the output voltage V_{od} to the input V_{id} is the differential voltage gain A_d of the differential amplifier. Therefore, combining Eqs. (1-9) and (1-10), we can show that

$$A_d = \frac{V_{od}}{V_{id}} = \frac{V_{C_1} - V_{C_2}}{V_1 - V_2}. \qquad\text{(1-11)}$$

The differential gain can be estimated with the following equation:

$$A_d \cong \frac{R_C}{r'_e + R_B/h_{fe}}, \qquad\text{(1-12a)}^*$$

---

*See Appendix B for derivation. The parameter r'_e is the dynamic (ac) resistance of the forward-biased emitter-base junction and is also referred to as h_{ib}. It is a hybrid parameter for the input resistance of a common emitter amplifier.

where R_C is the resistance in series with each collector,
    r'_e is the dynamic emitter-to-base resistance of each transistor,
    h_{fe} is the ac β, and
    R_B is the resistance seen looking to the left of either input to ground
    (the internal resistance of either input signal source V_1 or V_2).

Approximate values of r'_e are sometimes provided on manufacturers' data sheets, but can also be estimated with the equation

$$\frac{25\ mV}{I_C} \le r'_e \le \frac{50\ mV}{I_C}. \qquad (1\text{-}13)^\dagger$$

If the internal resistance R_B of each signal source is small compared to the transistors' r'_e, and if the h_{fe}'s are large, Eq. (1-12a) can be simplified to

$$A_d \cong \frac{R_C}{r'_e}. \qquad (1\text{-}12b)^*$$

If the collector resistors R_{C_1} and R_{C_2} are equal, the change in the output voltage at either collector to ground is half the change in the differential output voltage V_{od}. Therefore, the voltage gain A_v as a ratio of either collector voltage *change* measured to ground, to the differential input voltage *change* is half the differential voltage gain. Thus Eqs. (1-12a) and (1-12b) become

$$A_v = \frac{R_c}{2(r'_e + R_B/h_{fe})}, \qquad (1\text{-}12c)$$

and

$$A_v = \frac{R_c}{2r'_e}. \qquad (1\text{-}12d)$$

---

*See Appendix B for derivation. The parameter r'_e is the dynamic (ac) resistance of the forward-biased emitter-base junction and is also referred to as h_{ib}. It is a **hybrid** parameter for the input resistance of a common emitter amplifier.
†See Appendix C for derivation.

1.3   A BJT CURRENT SOURCE

The current source consisting of R_E and V_{EE} in the circuit of Fig. 1-4 is incapable of supplying a current that is constant enough for most applications of differential-input amplifiers. The advantage of a *very* constant current source is discussed later. For the present, be aware that the more constant the output of the current source, the better is the differential amplifier. A more sophisticated, and practical, current source is shown in Fig. 1-5. It consists of bias resistors R_1, R_2, and R_E, and the transistor Q_3.

   Note that R_1 and R_2 form a voltage divider that has the dc supply voltage $-V_{EE}$ applied to it. Note that the voltage across R_1 is the base-to-ground voltage V_B. Therefore,

$$V_B = \frac{-V_{EE}R_1}{R_1 + R_2}. \qquad (1\text{-}14)$$

The emitter-to-ground voltage is V_E and is equal to the V_B voltage less the drop across the forward-biased base-emitter junction; that is,

$$V_E = V_B - V_{BE} = \frac{-V_{EE}R_1}{R_1 + R_2} - V_{EE}. \qquad (1\text{-}15)$$

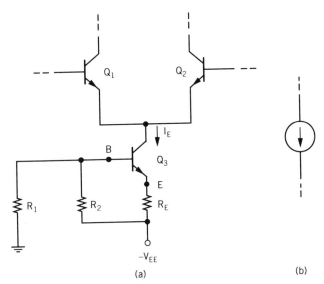

(a)   (b)

Figure 1-5   (a) BJT constant current source and (b) its symbol.

The voltage across R_E is the difference in the V_E voltage and the dc supply $-V_{EE}$. Thus

$$V_{R_s} = V_E - (-V_{EE}) = V_E + V_{EE}. \qquad (1\text{-}16)$$

The current through R_E is the collector current of Q_3. By Ohm's law, its value is

$$I_C \cong I_E = \frac{V_E + V_{EE}}{R_E}. \qquad (1\text{-}17)$$

Example 1-1

(a) In the circuit of Fig. 1-6, find the value of constant current I_E. Use $V_{BE} = 0.6$ V.

(b) What are the dc to ground voltages on collectors of Q_1 and Q_2 if $V_1 = V_2 = 0$ V?

(c) What is the differential output voltage V_{od}?

With Eq. (1-13), we can assume that on the average,

$$r'_e \cong \frac{30 \text{ mV}}{I_c}, \qquad (1\text{-}18)$$

where I_C is the collector of each BJT. Assume that

$$I_{C_1} = I_{C_2} = I_C = I_E/2,$$

where I_E is the output of the current source Q_3.

(d) What is the differential gain A_d of this circuit?

(e) If $V_1 = +15$ mV and $V_2 = -13$ mV, what are the differential input voltage V_{id} and the differential output voltage V_{od}?

Answers

(a) In this case, by Eq. (1-14)

$$V_B = \frac{-10 \text{ V} (4.7 \text{ k}\Omega)}{4.7 \text{ k}\Omega + 4.7 \text{ k}\Omega} = -5 \text{ V}.$$

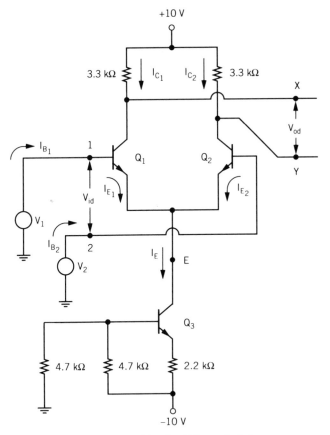

Figure 1-6   Circuit of Example 1-1.

Therefore,

$$V_E = -5 \text{ V} - 0.6 \text{ V} = -5.6 \text{ V}, \qquad (1\text{-}15)$$

and the voltage across the 2.2 kΩ is

$$V_{R_E} = -5.6 \text{ V} - (-10 \text{ V}) = 4.4 \text{ V}. \qquad (1\text{-}16)$$

The collector and emitter current of Q_3 is

$$I_C = I_E = \frac{4.4 \text{ V}}{2.2 \text{ k}\Omega} = 2 \text{ mA}. \qquad (1\text{-}17)$$

(b) Since the current source driving Q_1 and Q_1 is 2 mA and their base leads are at the same potential, each conducts 1 mA; that is, $I_{C_1} = I_{C_1} = 1$ mA.

By Eqs. (1-8a) and (1-8b),

$$V_{C_1} = V_{C_2} = 12 \text{ V} - 3.3 \text{ k}\Omega \,(1 \text{ mA}) = 8.7 \text{ V}.$$

(c) The differential output $V_{od} = 8.7 \text{ V} - 8.7 \text{ V} = 0 \text{ V}$.

(d) In this case

$$r'_e \cong \frac{30 \text{ mV}}{1 \text{ mA}} = 30 \,\Omega, \qquad (1\text{-}18)$$

and since $R_B = 0$, the gain

$$A_d = \frac{3{,}300 \,\Omega}{30 \,\Omega} = 110.$$

(e) Rearranging Eq. (1-11), we can show that

$$V_{od} = A_d V_{id} = 110 \,(15 \text{ mV} - (-13 \text{ mV})) = 3.08 \text{ V}.$$

The collector voltage V_{C_1} decreases by half of this V_{od} (by $3.08/2 = 1.54 \text{ V}$) and V_{C_2} increases by that same value; that is,

$$V_{C_1} = 8.7 \text{ V} - 1.54 \text{ V} = 7.16 \text{ V}$$

and

$$V_{C_2} = 8.7 \text{ V} + 1.54 \text{ V} = 10.24 \text{ V}.$$

The temperature stability of the current source of Fig. 1-5 is improved by the addition of two diodes as shown in Fig. 1-7. They serve to compensate for the changes in Q_3's base-emitter voltage V_{BE} with temperature changes. These series diodes are forward biased, as is the base-emitter junction. The drop across these diodes changes in the same direction as do changes in V_{BE}. Thus if, say, the temperature increases, causing V_{BE} to decrease, the drop across both diodes decreases proportionally.* This tends to keep the current I_E very constant. In other words, a decreased V_{BE} will increase both the drop across and the current I_E through resistor R_E. However, the decrease in the drop across the diodes occurring simultaneously will increase the drop across R_2. This causes the voltage across R_1, which is the voltage applied to the base V_B, to decrease. The resulting decrease in base current I_B opposes the increase in I_E. Of course, then, if the temperature decreases, which causes I_E to decrease, current I_B increases and again opposes the change in I_E.

*The forward-biased voltage drop of a silicon junction varies inversely with temperature by 2.5 mV/°C.

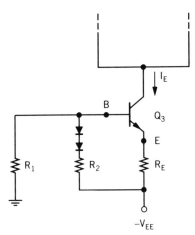

Figure 1-7   Temperature-stabilized current source.

1.4   CURRENT-SOURCED BIASED DIFFERENTIAL AMPLIFIER

A typical FET differential amplifier is shown in Fig. 1-8. It works in much the same way the BJT version does but has much larger resistance as seen looking into its input terminals 1 and 2. Integrated circuit differential and operational amplifiers are available with an FET-input stage for applications where extremely large input resistance is needed or where significant dc

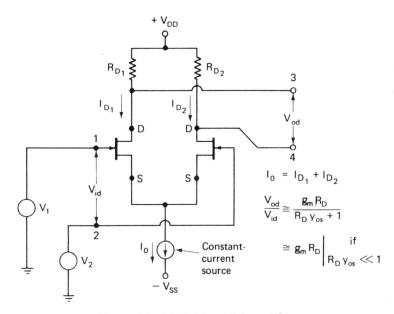

Figure 1-8   FET differential amplifier.

input bias currents are undesirable. The differential voltage gain of the FET stage can be estimated with the equation

$$A_d = \frac{V_{od}}{V_{id}} \cong \frac{g_m R_D}{R_D y_{os} + 1},$$

where g_m is the forward transductance,
$\quad\quad y_{os}$ is the output admittance, and
$\quad\quad R_D$ is the resistance in series with the drain of either FET.

Typically, the product $R_D y_{os} \ll 1$, and therefore the above can be simplified to

$$\boxed{A_d \cong g_m R_D.}$$

$$(1\text{-}19)^*$$

FET parameters g_m and y_{os} are provided on manufacturers' data sheets.

1.5   MULTISTAGE DIFFERENTIAL CIRCUITS

Cascading amplifier stages provides an overall (total) gain that is the product of the individual stage gains. In many applications, such as in linear ICs, differential amplifiers are cascaded for large total gain as shown in Fig. 1-9. A point to note is that the outputs of the first stage are *directly coupled* to the inputs of the second stage. This presents a problem that requires different biasing methods in these individual stages. For reasons that are discussed in detail in a later chapter, the input (first) stage must be able to take input voltages V_1 and V_2 that will vary positively or negatively with respect to ground and have as a result of these input variations, continuous variations of its differential output voltage V_{od}. This means that the BJTs or FETs of the input stage should not saturate or cut off with inputs in the neighborhood of about 0 V to ground or common. A BJT is saturated when the current into its base is so large as to reduce its V_{CE} voltage to about zero. It is cut off when its base current is essentially zero, causing the transistor to behave as an open between its collector and emitter.

The constant-current source in the input stage holds the emitters in Fig. 1-6 and the sources S in Fig. 1-8 at about ground potential. This allows use of input voltages that can be varied somewhat around ground potential because the bases must normally be at approximately the same voltage to ground as are the emitters. The second stage, however, has a considerable dc

---

*See Appendix D for derivation.

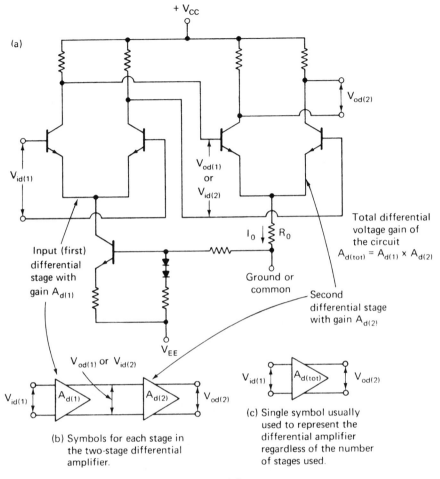

Figure 1-9

component applied to each of its inputs (bases) because the collectors of the first stage are above ground (have a dc voltage to ground) and are the inputs of the second stage. To avoid saturation of the transistors of the second stage, their emitters must be above ground by about the same potential as their bases. This is accomplished by the use of resistor R_0, one end of which is grounded instead of terminated to a negative dc bias source. The differential gain of this two-stage combination is

$$A_{d(\text{tot})} = \frac{V_{od(2)}}{V_{id(1)}} = A_{d(1)} \times A_{d(2)}.$$

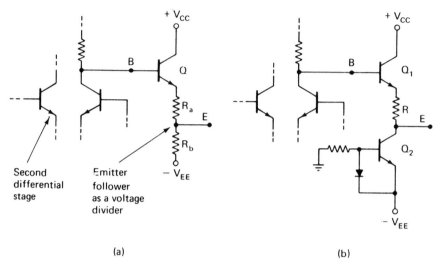

(a)   (b)

Figure 1-10   Level-shifting techniques.

1.6   LEVEL SHIFTING WITH INTERMEDIATE STAGE

All Op Amps have the equivalent of the cascaded differential amplifiers shown in Fig. 1-9. However, instead of two outputs, an Op Amp has a *single* output. *Ideally*, if $V_{id(1)} = 0$ V on the input (first) stage, an Op Amp's single output is 0 V to ground. If $V_{id(1)}$ is made slightly positive or negative, the Op Amp output swings to more negative or more positive values, respectively. In other words, the single output of an Op Amp typically varies between the dc supply voltages (the rails).

Since both outputs of the second stage in Fig. 1-9 are normally well above ground, neither of them will serve as the single Op Amp output terminal. However, either output can be shifted down to about 0 V to ground with appropriate voltage-divider circuitry. For example, an output of the second differential stage can work into an emitter-follower stage as shown in Fig. 1-10a. Thus a positive voltage at point B can cause 0 V to ground at point E with proper selection of components. That is, if the sum of the voltage drops across the transistor Q and resistor R_a equals the V_{CC} voltage, then the resistor R_b must drop the V_{EE} voltage, and point E is at 0 V with respect to ground. Better results are obtained by using a transistor stage Q_2 as a constant-current source, as shown in Fig. 1-10b, instead of the combination of R_b and V_{EE}. In this case then, point E is near ground potential though point B is well above ground. If point B is driven more positively by the preceding differential stage, transistor Q_1 will drop less voltage and Q_2 will drop more, causing point E to swing positively. And if point B swings

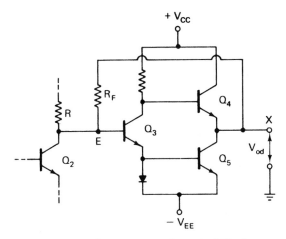

Figure 1-11   Output stage of a typical Op Amp.

negatively, Q_1's drop increases and Q_2's drop decreases, causing point E to swing negatively.

1.7   OUTPUT STAGE AND COMPLETE OPERATIONAL AMPLIFIER

Though the output terminal E of the level-shifting stage is not as positive as are the collectors of the preceding differential stage, this output E is not usually used as the output terminal of the Op Amp. An additional output stage, such as in Fig. 1-11, is used. The transistor Q_3 amplifies the signal at point E and drives the *totem-pole* arrangement of transistors Q_4 and Q_5. This output stage not only provides zero or near-zero voltage to ground at its output X when the differential input V_{id} to the first stage is zero, but also provides large peak-to-peak output signal capability when an input V_{id} is applied. In other words, the output X can swing positively to about the V_{CC} voltage and negatively to about the V_{EE} voltage. The resistor R_F provides feedback that stabilizes the output stage. It limits the drift of voltage at x with temperature changes.

A typical complete operational amplifier is shown in Fig. 1-12. By examining this circuit, we can see the stages discussed previously. While specific inner design of IC Op Amps is continually evolving, including FET or Darlington pair inputs, internal compensations, input voltage and output current limiter, and many other features, the circuits of this chapter give us a basic understanding of the inner operation of the IC Op Amp.

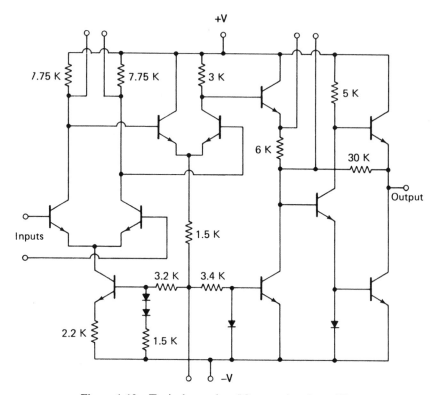

Figure 1-12   Typical complete IC operational amplifier.

REVIEW QUESTIONS 1

1-1. In which lead of a BJT is the current very much smaller than in the other two leads? Name the other two currents.

1-2. What is the symbol for the ratio of collector current to the base current?

1-3. What is the symbol for the dynamic resistance of a forward-biased base-emitter junction of a BJT?

1-4. Name the leads of a JFET and the currents that flow in them.

1-5. In amplifier applications, the base-emitter junction of a BJT is (*forward*) (*reverse*) biased and the gate-source junction is (*forward*) (*reverse*) biased. If the same voltage is applied to both inputs of an ideal differential amplifier, what will happen to the differential output voltage? Explain why.

1-6. In what way is a differential amplifier the same as an operational amplifier?

1-7. In what way is a differential amplifier different from an operational amplifier?

1-8. Over what approximate range can the output voltage of the Op Amp of Fig. 1-12 swing?

1-9. If the voltage at input 1 of Fig. 1-13 is driven a little more positive, output 3 will (*become more positive*), (*become more negative*), (*stay the same*) and the voltage at output 4 will (*become more positive*), *become more negative*), (*stay the same*).

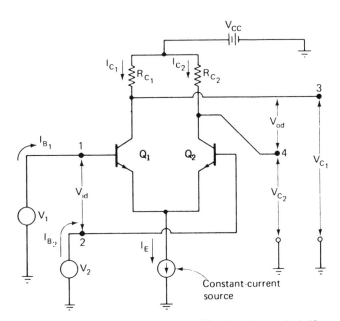

Figure 1-13   BJT differential amplifier, Problem 1-1–1-10.

1-10. What do the symbols h_{FE}, r'_e, and g_m represent?

PROBLEMS 1

Section 1.2

1-1. In the circuit of Fig. 1-13, if $V_{CC} = 15$ V, $V_1 = V_2 = 0$ V, $R_{C_1} = R_{C_2} = 3.9$ kΩ, and the constant current $I_E = 4$ mA, what are the values of V_{id}, V_{C_1}, V_{C_2}, and V_{od}?

1-2. In the circuit of Fig. 1-13, if $V_{CC} = 12$ V, $V_1 = V_2 = 0$ V, $R_{C_1} = R_{C_2} = 4.7$ kΩ, and the constant current $I_E = 2$ mA, what are the values of V_{id}, V_{C_1}, V_{C_2}, and V_{ov}?

1-3. In the circuit of Fig. 1-13, if $V_{CC} = 15$ V, $R_{C_1} = R_{C_2} = 3.9$ kΩ, the constant current $I_E = 4$ mA, and V_{C_1} measures 8 V dc, what are the values of I_{C_1}, I_{C_2}, V_{C_2}, and V_{od}?

1-4. In the circuit of Fig. 1-13, if $V_{CC} = 12$ V, $R_{C_1} = R_{C_2} = 4.7$ kΩ, the constant current $I_E = 2$ mA, and V_{C_2} measures 8.3 V dc, what are the values of I_{C_1}, I_{C_2}, V_{C_1}, and V_{od}?

1-5. Referring to the circuit described in Problem 1-1, what is the value of V_{C_1} if input 1 becomes open from ground?

1-6. Referring to the circuit described in Problem 1-2, what are the values of V_{C_1} and V_{C_2} if input 2 becomes open from ground?

1-7. In the circuit of Fig. 1-13, if $R_{C_1} = R_{C_2} = 3.9$ kΩ and the constant current $I_E = 4$ mA, what are the voltage gains (a) A_d and (b) A_v? Assume that the internal resistances of the input signal sources are negligible and that $r'_e = 30$ mV$/I_C$.

1-8. In the circuit of Fig. 1-13, if $R_{C_1} = R_{C_2} = 4.7$ kΩ and the constant current $I_E = 2$ mA, what is the voltage gain V_{od}/V_{id}? Assume that the internal resistances of the input signal sources are negligible and that $r'_e = 30$ mV$/I_C$.

1-9. In the circuit of Fig. 1-13, if $V_{CC} = 15$ V, $V_1 = 13$ mV dc, $V_2 = -12$ mV dc, $R_{C_1} = R_{C_2} = 3.9$ kΩ, and the constant current $I_E = 4$ mA, what are the values of V_{id}, V_{C_1}, V_{C_2}, and V_{od}? Hint: See (e) of Example 1-1.

1-10. In the circuit of Fig. 1-13, if $V_{CC} = 12$ V, $V_1 = 15$ mV dc, $V_2 = 25$ mV dc, $R_{C_1} = R_{C_2} = 4.7$ kΩ, and the constant current $I_E = 2$ mA, what are the values of V_{id}, V_{C_1}, V_{C_2}, and V_{od}?

Section 1.3

1-11. In the circuit of Fig. 1-14, if $V_{BE} = 0.675$ V, what is the value of current I_0?

1-12. Replace the 4.7-kΩ resistor in the circuit of Fig. 1-14 with 2.7 kΩ. With $V_{BE} = 0.7$ V, what is the resulting I_0?

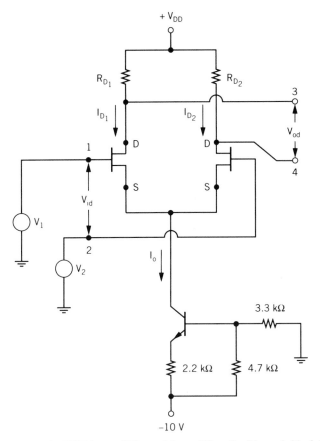

Figure 1-14 FET-input differential amplifier, Problems 1-11–1-14.

1-13. If in the circuit of Fig. 1-14, $V_1 = V_2 = 1.5$ V, $R_{D_1} = R_{D_2} = 2.2$ kΩ, and $V_{DD} = 9$ V, what are the drain-to-ground voltages V_{D_1} and V_{D_2} and the differential output V_{od}?

1-14. Replace the 4.7-kΩ resistor in the circuit of Fig. 1-14 with 2.7 kΩ. With $V_1 = V_2 = -0.5$ V, $V_{DD} = 9$ V and, $R_{D_1} = R_{D_2} = 2.7$ kΩ, what are the values of V_{id}, the drain-to-ground voltages V_{D_1} and V_{D_2}, and the differential output V_{od}?

2

CHARACTERISTICS OF OP AMPS AND THEIR POWER SUPPLY REQUIREMENTS

The term *operational amplifier* refers to a high-gain dc amplifier that has a differential input (two input leads) and a single-ended output (one output lead). The signal output voltage V_o is larger than the differential input signal across the two inputs by the gain factor of the amplifier (see Fig. 2-1). Op Amps have characteristics such as high input resistance, low output resistance, high gain, etc., that make them highly suitable for many applications, a number of which are shown and discussed in later chapters. By examining some applications and comparing the characteristics of typical Op Amps, we will see that some types are apparently better than others. The overall most desirable characteristics that Op Amp manufacturers strive to obtain in their products are *ideal characteristics*. While some of these ideal characteristics are impossible to obtain, we often assume that they exist to develop circuits and equations that work perfectly on paper. These circuits and equations then work very well with practical Op Amps. Actual Op Amp characteristics are more or less ideal, relative to conditions external to the Op Amp: signal source resistance, load resistance, amount of feedback used, etc. Some of the more important characteristics, ideal and practical values, are given and defined in the following sections.

2.1 OPEN-LOOP VOLTAGE GAIN* A_{VOL}

The open-loop voltage gain A_{VOL} of an Op Amp is its *differential* gain under conditions where no negative feedback is used, as shown in Figs. 2-1 and 2-2.

---

*In industry, the open-loop voltage gain is referred to with a variety of symbols: A_{VOL}, A_d, A_{EOL}, A_{VD}, etc.

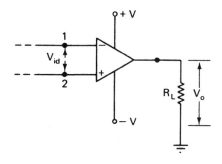

Figure 2-1 Typical Op Amp symbol. $+V$ is the positive dc power supply voltage; $-V$ is the negative dc power supply voltage; $A_{VOL} = V_o/V_{id}$.

Ideally, its value is infinite—that is, the ideal Op Amp has an open-loop voltage gain

$$A_{VOL} = \frac{V_o}{V_{id}} = -\infty \qquad (2\text{-}1a)$$

or

$$A_{VOL} = \frac{V_o}{V_1 - V_2} = -\infty \qquad (2\text{-}1b)$$

The negative sign means that the output V_o and the input V_{id} are out of phase. Of course, the concept of an infinite gain is theoretical. The important point to understand is that the Op Amp's output signal voltage V_o should be very much larger than its differential input signal V_{id}. To put it another way, the input V_{id} should be infinitesimal compared to any practical value of output V_o. Typical open-loop gains A_{VOL} range from about 5000 (about 74 dB) to 1,000,000 (about 120 dB)—that is,

$$5000 \le A_{VOL} \le 1{,}000{,}000 \qquad (2\text{-}2a)$$

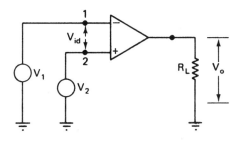

Figure 2-2 Op Amp with input voltages V_1 and V_2 whose difference is V_{id}. Power supply voltages $+V$ and $-V$ are assumed if not shown.

or

$$74 \text{ dB} \le A_{VOL} \le 120 \text{ dB} \qquad (2\text{-}2\text{b})$$

with popular types of Op Amps. The fact that the output signal V_o is 5000 to 10^6 times larger than the differential input signal V_{id} does not mean that V_o can actually be very large. In fact, its positive and negative peaks are limited to values a little less than the positive and negative supply voltages being used to power the Op Amp. Since the dc supply voltages of monolithic IC Op Amps are usually less than 20 V, the peaks of their output voltages V_o are less than 20 V. This fact, coupled with the large A_{VOL} of the typical Op Amp, makes the voltage V_{id} across its inputs 1 and 2 very small. Of course, the larger the open-loop gain A_{VOL}, the smaller V_{id} is in comparison to any practical value of output signal V_o. Thus, since

$$A_{VOL} = \frac{V_o}{V_{id}} \qquad (2\text{-}3\text{a})$$

or

$$A_{VOL} = \frac{V_o}{V_1 - V_2}, \qquad (2\text{-}3\text{b})$$

then

$$V_{id} = \frac{V_o}{A_{VOL}}$$

and, therefore,*

$$\lim V_{id} = 0,$$
$$A_{VOL} \to \infty.$$

Thus, if the gain A_{VOL} in the expression V_o/A_{VOL} is very large, the value of this expression, which is V_{id}, must be very small. In fact, V_{id} is usually so small that we can assume there is practically no potential difference between the inverting $(-)$ and noninverting $(+)$ inputs.

---

*As the value of open-loop gain A_{VOL} approaches infinity, the value of differential input voltage V_{id} approaches zero if $V_o \ne 0$.

2.2 OUTPUT OFFSET VOLTAGE V_{oo}

The output offset voltage V_{oo} of an Op Amp is its output voltage to ground or common under conditions when its differential input voltage $V_{id} = 0$ V. Ideally $V_{oo} = 0$ V. In practice, due to imbalances and inequalities in the differential amplifiers within the Op Amp itself (see Fig. 1-12), some output offset voltage V_{oo} will usually occur even though the input $V_{id} = 0$ V. In fact, if the open-loop gain A_{VOL} of the Op Amp is high and if no feedback is used, the output offset is large enough to saturate the output. In such cases, the output voltage is either a little less than the positive source voltage $+V$ or the negative source $-V$. This is not as serious as it first appears because corrective measures are fairly simple to apply and because the Op Amp is seldom used without some feedback. When corrective action is taken to bring the output to 0 V, when the applied input signal is 0 V, the Op Amp is said to be *balanced* or *nulled*.

If both inputs are at the same finite potential, causing $V_1 - V_2 = V_{id} = 0$ V, and if the output $V_o = 0$ V as a result, the Op Amp is said to have an ideal *common-mode rejection* (CMR). Though most practical Op Amps have a good CMR capability, they do pass some *common-mode* (CM) voltage to the load R_L. Typically though, the load's common-mode output voltage V_{cmo} is hundreds or thousands of times *smaller* than the input common-mode voltage V_{cm}. A thorough discussion of the practical Op Amp's CMR capability and its applications is given in a later chapter.

2.3 INPUT RESISTANCE R_i

The Op Amp's input resistance R_i is the resistance seen looking into its inputs 1 and 2 as shown in Fig. 2-3. Ideally, $R_i = \infty \, \Omega$. Practical Op Amps' input resistances are not infinite but instead range from less than 5 kΩ to over 20 MΩ, depending on type. Though resistances in the low end of this range seem a bit small compared to the desired ideal ($\infty \, \Omega$), they can be quite large compared to the low internal resistances of some signal sources that commonly drive the inputs of Op Amps. Generally, if a high-resistance signal source is to drive an Op Amp, the Op Amp's input resistance should be relatively large. As we will see later, the *effective* input resistance $R_{i(\text{eff})}$ is made considerably larger than the manufacturer's specified R_i by use of

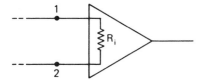

Figure 2-3 Input resistance R_i is the resistance seen looking into the inputs 1 and 2.

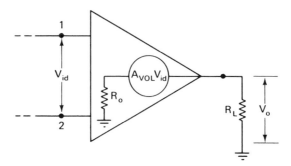

Figure 2-4   Output resistance R_o tends to reduce output voltage V_o.

negative feedback. Manufacturers usually specify their Op amps' input resistances as measured under open-loop conditions (no feedback). In most linear applications, Op Amps are wired with feedback and this improves (increases) the effective resistance seen by the signal source driving the Op Amp.

2.4   OUTPUT RESISTANCE R_o

With a differential input signal V_{id} applied, the Op Amp behaves like a signal generator as the load connected to the output sees it. As shown in Fig. 2-4, the Op Amp is equivalent to a signal source generating an open-circuit voltage of $A_{VOL}V_{id}$ and having an internal resistance of R_o. This R_o is the Op Amp's output resistance and ideally should be 0 Ω. Obviously, if $R_o = 0$ Ω in the circuit of Fig. 2-4, all of the generated output signal $A_{VOL}V_{id}$ appears at the output and across the load R_L. Depending on the type of Op Amp, specified output resistance R_o values range from a few ohms to a few hundred ohms and are usually measured under open-loop (no feedback) conditions. Fortunately, the *effective* output resistance $R_{o(eff)}$ is considerably smaller when the Op Amp is used with feedback. In fact, in most applications using feedback, the effective output resistance of the Op Amp is very nearly ideal (0 Ω). The effect feedback has on output resistance is discussed in more specific terms later.

2.5   BANDWIDTH BW

The *bandwidth* BW of an amplifier is defined as the range of frequencies at which the output voltage does not drop more than 0.707 of its maximum value while the input voltage *amplitude* is constant. Ideally, an Op Amp's bandwidth BW = ∞. An infinite bandwidth is one that starts with dc and extends to infinite cycles/second (Hz). It is indeed idealistic to expect such a bandwidth from any kind of amplifier. Practical Op Amps fall far short of this ideal. In fact, limited high-frequency response is a shortcoming of monolithic IC Op Amps. While some types of Op amps can be used to amplify signals up to a few megahertz, they require carefully chosen feedback and externally wired compensating components. Most general-purpose IC Op Amps are limited to less than 1-MHz bandwidth and more often are used with signals well under a few kilohertz, especially if any significant gain is expected of them. The limited high-frequency response of Op Amps is not a serious matter with most types of instrumentation in which Op Amps are used extensively. Chapter 6 discusses high frequency characteristics of Op Amps.

2.6 RESPONSE TIME

The response time of an amplifier is the time it takes the output of an amplifier to change after the input voltage changes. Ideally, response time = 0 seconds; that is, the output voltage should respond instantly to any change on the input. Figure 2-5 shows an Op Amp's typical output-voltage response to a step input voltage when wired for unity gain. Manufacturers also specify a *slew rate* that gives the circuit designer a good idea of how fast a given Op Amp responds to changes of input voltage. Note that the time scale in Fig. 2-5 is in microseconds (μs) and that the output voltage, when changing, overshoots the level it eventually settles at. Overshoot is the ratio of the amount of overshoot to the steady-state deviation expressed as a percentage. For example, if we observe the responding output voltage of Fig. 2-5 on an oscilloscope whose vertical deflection sensitivity is at 10 V/cm, and the amount of overshoot measures 0.2 cm, the percentage of overshoot is

$$\frac{\text{Amount of overshoot}}{\text{Steady-state deflection}} \times 100 = \frac{0.2 \text{ cm}}{2 \text{ cm}} \times 100 = 10\%.$$

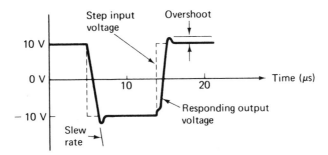

Figure 2-5 Typical Op Amp response to a step input voltage when wired as a voltage follower (unity gain).

From time to time we will refer to the various *ideal* characteristics, and here is a summary:

(1) Open-loop voltage gain $A_{VOL} = \infty$.
(2) Output offset voltage $V_{oo} = 0$ V.
(3) Input resistance $R_i = \infty\ \Omega$.
(4) Output resistance $R_o = 0\ \Omega$.
(5) Bandwidth BW $= \infty$ Hz.
(6) Response time $= 0$ s.

2.7   POWER SUPPLY REQUIREMENTS

In many applications, the Op Amp's output voltage V_o must be capable of swinging in both positive and negative directions. In such applications, the Op Amp requires two source voltages: one positive $(+V)$ and the other negative $(-V)$ with respect to ground or a common point. These dc source voltages must be well filtered and regulated; otherwise the Op amp's output voltage will vary with the power supply variations. The output voltage of an Op Amp varies more or less with power supply variations, depending on its closed-loop* voltage gain and sensitivity factor S which is usually specified on the manufacturer's data sheets. An ideal Op Amp has a sensitivity factor $S = 0$, which means that power supply voltage variations have no effect on its output. Practical Op Amps, however, are affected by changes in the supply voltages, and therefore regulated supplies are used to keep Op Amp outputs responsive to differential input voltages only.

REVIEW QUESTIONS 2

Fill in the blanks of the next 10 problems with one of the 6 following terms:

(I)	infinite	(IV)	relatively small
(II)	zero	(V)	decrease
(III)	relatively large	(VI)	increase

2-1. The ideal Op Amp's output resistance is _____ ohms.

2-2. The ideal Op Amp's open-loop voltage gain is _____ .

2-3. The ideal input resistance of an Op Amp is _____ ohms.

2-4. The ideal Op Amp's bandwidth is _____ hertz.

2-5. The ideal response time of an Op amp is _____ seconds.

---

*The closed-loop gain of an Op Amp is its gain when a feedback loop is used.

2-6. Compared to the internal resistance of the signal source, the practical Op Amp's effective input resistance should be _____.

2-7. Compared to the load resistance, the effective output resistance of a practical Op Amp should be _____.

2-8. When the differential input voltage is zero, the Op Amp's output voltage ideally should be _____.

2-9. Feedback tends to _____ the effective output resistance, and to _____ the effective input resistance.

2-10. Compared to a typical output signal voltage, the differential input should ideally be _____.

2-11. Comparatively, what do the terms *open-loop voltage gain* and *closed-loop voltage gain* mean referring to Op Amps?

2-12. To which of the Op Amp characteristics does the term *slew rate* apply?

2-13. If the Op amp's output must be capable of swinging in both positive and negative directions, generally what kind of power supply does it require?

2-14. What does it mean when we say an Op Amp is *nulled*?

2-15. What is the meaning of the term CMR?

2-16. What undesirable effect might result if an Op Amp's dc source voltages are not well regulated?

3

THE OP AMP WITH
AND WITHOUT FEEDBACK

The available open-loop gain of an Op Amp is never used in linear applications. It is too large. If we tried to use the available A_{VOL}, even very small differential input signals V_{id} will drive an Op Amp's output into its rails causing severe clipping (distortion) of the resulting output V_o. In this chapter we will see how negative feedback is used with an Op Amp to reduce and stabilize its effective voltage gain. This effective voltage gain is called the *closed-loop* gain A_v. In fact, with feedback we are able to select any specific gain we need as long as it is less than the Op Amp's open-loop gain A_{VOL}.

3.1 OPEN-LOOP CONSIDERATIONS

The Op Amp is generally classified as a linear device. This means that its output voltage V_o tends to proportionally follow changes in the applied differential input V_{id}. Within limits, the changes in output voltage V_o are larger than the changes in the input V_{id} by the open-loop gain A_{VOL} of the Op Amp. The amount that the output voltage V_o can change (swing), however, is limited by the dc supply voltages and the load resistance R_L. Generally, the output voltage swing is restricted to values between the rails. As shown in Fig. 3-1, manufacturers provide curves showing their Op Amps' maximum output voltage swing versus supply voltage and also versus load resistance R_L. Note that smaller supply voltages and load resistances reduce an Op amp's signal output capability. Any attempt to drive the Op Amp beyond its limits results in sharp clipping of its output signals.

For example, the Op Amp in Fig. 3-2 is shown with a small input signal V_{id} applied to its inverting input 1 and its resulting output V_o. Note that the

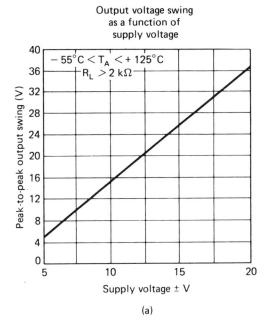

(a)

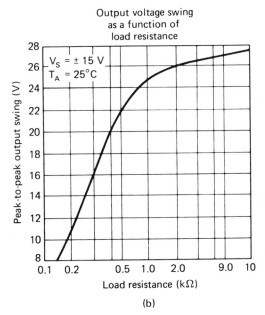

(b)

Figure 3-1 Typical Op Amp characteristics: (a) output voltage V_o swing vs. dc supply voltage, (b) output voltage V_o swing vs. load resistance R_L.

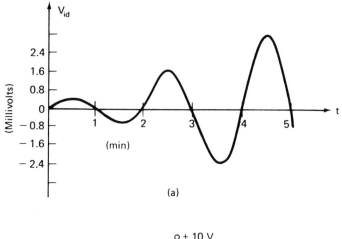

(a)

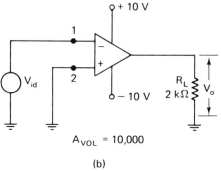

$A_{VOL} = 10,000$

(b)

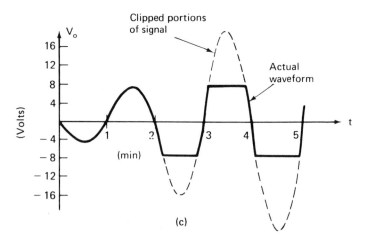

(c)

Figure 3-2 (a) Differential input signal, (b) Op Amp with no feedback (inverting mode), (c) output signal waveform if $A_{VOL} = 10^4$.

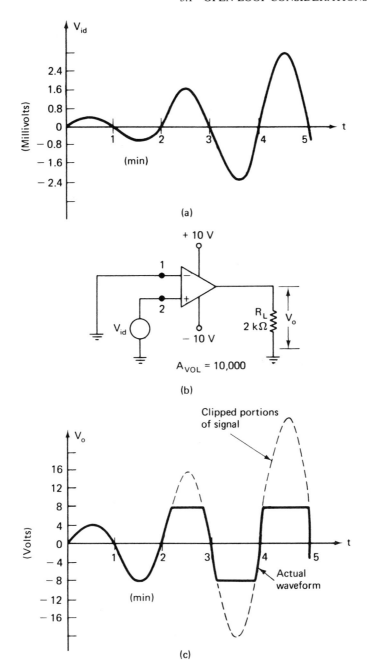

Figure 3-3   (a) Differential input signal, (b) noninverting Op Amp circuit with no feedback, (c) output signal waveform.

output voltage $V_o = 0$ V at the times when $V_{id} = 0$ V, that is, the circuit is nulled. (For reasons we will see later, it is difficult to null the output of an Op Amp if no feedback is used.) Note also that the gain $A_{VOL} = 10,000$, which means that the output signal V_o is -10^4 times larger than the input V_{id}, but within limits. In this case, the limits are the rails, plus and minus 8 V, as indicated by the fact that the output signal V_o is clipped at $+8$ V on some positive alternations and at -8 V on some negative alternations. Apparently, this Op Amp has output voltage swing vs. supply voltage characteristics as shown in Fig. 3-1a. Note that, although the input V_{id} varies somewhat sinusoidally and with relatively low amplitudes, the output V_o is -10^4 times larger and runs into the positive and negative rails. Consequently severe clipping occurs. Only the smaller input signals—1.6 mV peak to peak or less —are amplified without clipping. There are applications where the Op Amp is purposely used as a signal-squaring circuit in which deliberately driving its input with a relatively large signal results in a square-wave output. This output appears more square and less like a trapezoid when the input signals have larger amplitudes.

If the same input signal is applied to the noninverting input 2, as shown in Fig. 3-3, the output is $+10^4$ times larger, within limits of course. In this case, the output swings positively and then negatively on the positive and negative alternations of the input signal, respectively. Clipping occurs when the output attempts to exceed the Op Amp's rails, just as in the inverting mode.

Example 3-1

Sketch output voltage waveforms of an Op Amp, such as in Fig. 3-2, if the input signal is as shown but with:

(a) an Open-loop gain $A_{VOL} = 5000$.
(b) an Open-loop gain $A_{VOL} = 100,000$.

Answer. See Fig. 3-4. Note that with a larger gain, A_{VOL}, a given input signal tends to be more distorted. This does not mean that a high open-loop gain is undesirable; on the contrary, the larger A_{VOL} is, the better. As we will see, clipping can easily be controlled with feedback.

3.2   FEEDBACK AND THE INVERTING AMPLIFIER

Negative feedback is used with Op Amps in linear applications. Feedback enables the Op Amp circuit designer to easily select and control voltage gain. Generally, an amplifier has negative feedback if a portion of its output is fed

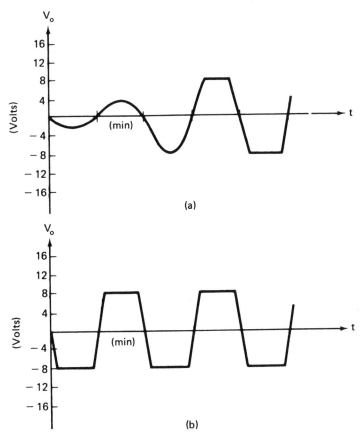

Figure 3-4   (a) Output voltage of the Op amp in Fig. 3-2 if $A_{VOL} = 5000$, (b) output voltage of the Op Amp in Fig. 3-2 if $A_{VOL} = 100,000$.

back to its inverting input. The Op amp circuit in Fig. 3-5 has negative feedback. Note that a resistor R_F is across the Op Amp's output and inverting input terminals. This, along with R_1, causes a portion of the output signal V_o to be fed back to the inverting input terminal. With this feedback, the circuit's effective (closed-loop) voltage gain A_v is typically much smaller than the Op Amp's open-loop gain A_{VOL}.

In the circuit shown in Fig. 3-5, the input signal is voltage V_s; therefore, its closed-loop voltage gain is

$$A_v = \frac{V_o}{V_s}. \tag{3-1}$$

This ratio is called the *closed-loop gain* because it is the gain when resistors

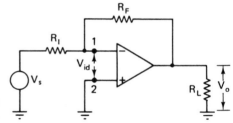

Figure 3-5 Op Amp connected to work as an inverting amplifier with feedback.

R_F and R_1 complete a loop from the amplifier's output to its inverting input 1. The specific value of A_v is determined mainly by the values of resistors R_F and R_1.

We can see how resistors R_1 and R_F in the circuit of Fig. 3-5 determine the closed-loop gain A_v if we analyze this circuit's current paths and voltage drops. For example, as shown in Fig. 3-6, the signal source V_s drives a current I through R_1. Assuming that the Op Amp is ideal (having infinite resistance looking into input 1), all of this current I flows up through R_F. By Ohm's law we can show that the voltage drop across R_1 is $R_1 I$ and that the voltage across R_F is $R_F I$. The significance of this will be seen shortly.

Since the open-loop gain A_{VOL} of the Op Amp is ideally infinite or at least very large practically, the differential input voltage V_{id} is infinitesimal compared to the output voltage V_o. Thus, since

$$A_{VOL} = \frac{V_o}{V_{id}}, \tag{2-3a}$$

then

$$V_{id} = \frac{V_o}{V_{VOL}}.$$

This last equation shows that the larger A_{VOL} is with any given V_o, the

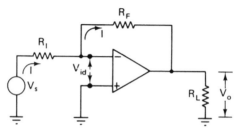

Figure 3-6 Currents in R_1 and R_F are equal if the Op Amp is ideal; they are approximately equal with practical Op Amp if R_F is not too large.

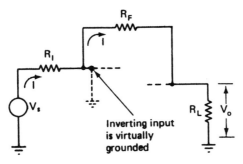

Figure 3-7 Equivalent circuit of Figs. 3-5 and 3-6; $V_s \cong R_1 I$ and $V_o \cong R_F I$.

smaller V_{id} must be compared to V_o. The point is, the voltage V_{id} across inputs 1 and 2 is practically zero because of the large open-loop gain A_{VOL} of the typical Op Amp. Therefore, with *virtually* no potential difference between inputs 1 and 2, and with the noninverting input 2 grounded, the inverting input is *virtually* grounded too. This circuit's equivalent can therefore be shown as in Fig. 3-7. With voltage V_o to ground at the right of R_F and virtual ground at its left, the voltage across R_F is V_o for practical purposes. By Ohm's law,

$$V_o \cong R_F I. \tag{3-2}$$

Similarly, we can see that the voltage V_s to ground is applied to the left of R_1, while its right end is virtually grounded. Therefore, for most practical purposes, the voltage across R_1 is V_s. Again with Ohm's law we can show that

$$V_s \cong R_1 I. \tag{3-3}$$

Since the closed-loop gain A_v is the ratio of the output signal voltage V_o to the input signal voltage V_s, we can substitute Eqs. (3-2) and (3-3) into this ratio and show that

$$A_v = \frac{V_o}{V_s} \cong -\frac{R_F I}{R_1 I} \cong -\frac{R_F}{R_1}. \tag{3-4}$$

The negative sign means that the input and output signals are out of phase. This last equation shows that by selecting a ratio of feedback resistance R_F

to the input resistance R_1, we select the inverting amplifier's closed-loop gain A_v.

There are limits on the usable values of R_F and R_1 which are caused by practical design problems. Though these design problems are discussed in later chapters, for the present we should know that the feedback resistance R_F is rarely larger than 10 MΩ. More frequently, R_F is 1 MΩ or less. Since one end of the signal source V_s is grounded and one end of R_1 is virtually grounded, V_s sees R_1 as the amplifier's input resistance. To avoid loading of (excessive current drain from) the signal source V_s, R_1 is typically 1 kΩ or more.

Example 3-2

Referring to the circuit in Fig. 3-8a, find its voltage gain V_o/V_s when the switch S is in:

(a) position 1,
(b) position 2, and
(c) position 3.
(d) If the switch is in position 2 and the input voltage V_s has the waveform shown in Fig. 3-8b, sketch the output voltage waveform V_o. Assume that when V_s was zero, the output V_o was zero (nulled).
(e) What is the resistance seen by the signal source V_s for each of the switch positions?

Answers

(a) When the switch S is in position 1, the feedback resistance $R_F = 10$ kΩ. The input resistor $R_1 = 1$ kΩ regardless of the switch position. Therefore, the gain is

$$A_v \cong -\frac{R_F}{R_1} = -\frac{10 \text{ k}\Omega}{1 \text{ k}\Omega} = -10. \qquad (3\text{-}4)$$

(b) With the switch S in position 2, $R_F = 100$ kΩ; therefore

$$A_v \cong -\frac{100 \text{ k}\Omega}{1 \text{ k}\Omega} = -100.$$

(c) With the switch S in position 3, $R_F = 1$ MΩ; therefore,

$$A_v \cong -\frac{1 \text{ M}\Omega}{1 \text{ k}\Omega} = -1000.$$

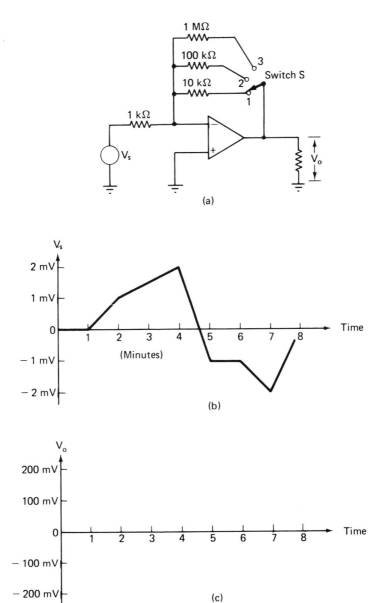

Figure 3-8 Circuit for Example 3-2.

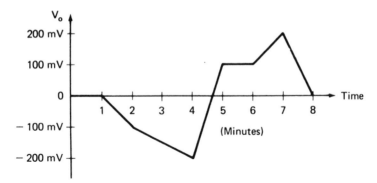

Figure 3-9 Answer to Example 3-2, part (d).

The negative signs mean that the input and output signal voltages are out of phase.

(d) Since the gain $A_v = -100$ with the switch S in position 2, the input signal voltage V_o is amplified by 100 and its phase is inverted, resulting in an output voltage waveform shown in Fig. 3-9. By rearranging Eq. (3-1), we can show that

$$V_o = A_v V_s.$$

Thus, at it $= 2$ min, $V_o = -100 \ (1 \ \text{mV}) = -100$ mV. At $t = 4$ min, $V_o = -100 \ (2 \ \text{mV}) = -200$ mV, etc.

(e) The signal source V_s sees R_1 as the load regardless of the switch position.

Example 3-3

The Op Amp in Fig. 3-8 operates with dc supply voltages of ± 15 V and with a load resistance $R_L = 200 \ \Omega$. If the characteristics shown in Fig. 3-1 are typical of this Op Amp, what are the values of:

(a) the maximum possible peak-to-peak unclipped output signal V_o, and
(b) the maximum input peak-to-peak signal voltage V_s that can be applied and not cause clipping of V_o while the switch S is in position 2?

Answers

(a) Referring to Fig. 3-1b, we project up from 0.2 kΩ to the curve. Directly to the left of the intersection of our projection and curve, we see that this amplifier's peak-to-peak output swing capability is about 11 V. This means that clipping will occur if we attempt to drive the output V_o beyond $+5.5$ V or -5.5 V.

(b) With the switch S in position 2, the gain is -100 as we found in the previous problem. Since the maximum peak-to-peak output swing is about 11 V, the maximum peak-to-peak input swing is determined as follows. Since

$$A_v = \frac{V_o}{V_s},$$ (3-1)

then

$$V_s = \frac{V_o}{V_v} \cong \frac{11 \text{ V (p-p)}}{-100} = |110 \text{ mV (p-p)}|.$$

3.3 FEEDBACK AND THE NONINVERTING AMPLIFIER

We can use the Op Amp, with feedback, in a noninverting mode much as we used it in the inverting mode discussed in the previous section. We can drive the noninverting input 2 with a signal source V_s instead of indirectly driving the inverting input 1 as shown in Figs. 3-3 and 3-10. As with the inverting amplifier, the values of externally connected resistors R_1 and R_F determine the circuit's closed-loop voltage gain A_v. A gain equation in terms of R_1 and R_F can be worked out if we analyze the noninverting amplifier's currents and voltages.

As shown in Fig. 3-11a, the signal current I is the same through resistors R_1 and R_F as long as the resistance R_i looking into the inverting input 1 is infinite or at least very large. In other words, R_1 and R_F are effectively in series. As before, the voltge drops across these resistors can be shown as $R_1 I$ and $R_F I$. At the right side of R_F we have the output voltage V_o to ground. This voltage is also across R_1 and R_F because these resistors are effectively in series and the left side of R_1 is grounded. Thus, we can show that the output voltage V_o is the sum of the drops across R_1 and R_F; that is

$$V_o \cong R_1 I + R_F I.$$

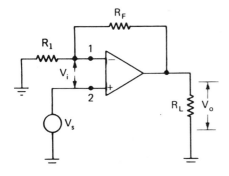

Figure 3-10 Op Amp wired to work as a noninverting amplifier (noninverting mode).

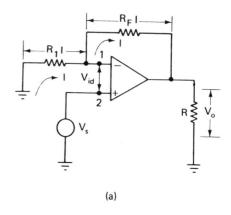

(a)

Figure 3-11 (a) Noninverting amplifier with currents and voltage drops shown; (b) the difference voltage V_{id} across inputs 1 and 2 is so small compared to V_s that these inputs are virtually shorted.

Inputs virtually at the same potential

(b)

Factoring I out, we get

$$V_o \cong (R_1 + R_F)I. \qquad (3\text{-}5)$$

As with the inverting amplifier, the differential input voltage V_{id} is zero for most practical purposes. Due to the very large open-loop gain A_{VOL} of the typical Op Amp, we can assume that there is practically no potential difference between points 1 and 2 as shown in Fig. 3-11b. Therefore, nearly all of the input signal voltage V_s appears across R_1, and by Ohm's law, we can again show that

$$V_s \cong R_1 I. \qquad (3\text{-}3)$$

The voltage gain A_v of the noninverting amplifier is the ratio of its output V_o to its input V_s. Thus, substituting the right sides of Eqs. (3-3) and (3-5) into this ratio yields

$$A_v = \frac{V_o}{V_s} \cong \frac{(R_F + R_1)I}{R_1 I} = \frac{R_F + R_1}{R_1} = \frac{R_F}{R_1} + 1. \qquad (3\text{-}6)$$

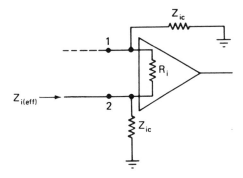

Figure 3-12   Equivalent circuit of an Op Amp; R_i is the resistance between inputs 1 and 2 specified by the manufacturer; Z_{ic} is the impedance to ground or common from either input which is actually internal to the Op Amp.

Since the resistance seen looking into the noninverting input is infinite if the Op Amp is ideal (or at least very large with a practical Op Amp), the signal source V_s sees a very large resistance looking into the Op Amp of Fig. 3-10. Generally, the effective input resistance of the noninverting amplifier is larger with circuits wired to have lower closed-loop gains. More specifically, the effective opposition to signal current flow into the noninverting input is an impedance $Z_{i(eff)}$ when input capacitances are considered. Looking into the noninverting input 2, the signal current sees parallel paths: through an impedance Z_{ic} and through the Op Amp's resistance R_i (see Fig. 3-12). Thus the total effective input resistance is

$$R_{i(eff)} \cong \frac{1}{A_v/A_{VOL}R_i + 1/Z_{ic}} \qquad (3\text{-}7a)$$

where R_i is the input resistance measured between input 1 and 2, open-loop, and its value is provided on the manufacturer's spec sheets,

Z_{ic} is the impedance to ground or common measured from either input,

A_{VOL} is the open-loop gain, and

A_v is the closed-loop gain.

Typically, Z_{ic} is much larger than R_i, especially at lower frequencies. If we assume that Z_{ic} is relatively very large, Eq. (3-7a) can be simplified to

$$R_{i(eff)} \cong \frac{1}{A_v/A_{VOL}R_i} = \left(\frac{A_{VOL}}{A_v}\right)R_i. \qquad (3\text{-}7b)$$

Note that if the Op Amp is wired to have a low closed-loop gain A_v, say approaching 1, the effective input resistance $R_{i(eff)}$ theoretically approaches a value that is A_{VOL} times larger than the manufacturer's specified R_i.

Example 3-4

If $R_1 = 1\ \text{k}\Omega$ and $R_F = 10\ \text{k}\Omega$ in the circuit of Fig. 3-10, what is the voltage gain V_o/V_s? If this circuit's input signal voltage V_s has the waveform shown in Fig. 3-13a, sketch the resulting output V_o. Assume that the Op Amp was initially nulled.

Answer. Since this Op Amp is wired in a noninverting mode, the voltage gain can be determined with Eq. (3-6):

$$A_v \cong \frac{R_F}{R_1} + 1 = 11.$$

Therefore, the output voltage is 11 times larger than, and in phase with, the input signal V_s. Thus at $t = 2$ min, $V_o = 11(-20\ \text{mV}) = -220\ \text{mV}$. Similarly, at $t = 5$ min, $V_o = 11(10\ \text{mV}) = 110\ \text{mV}$, etc. (See Fig. 3-14 for the complete output waveform.)

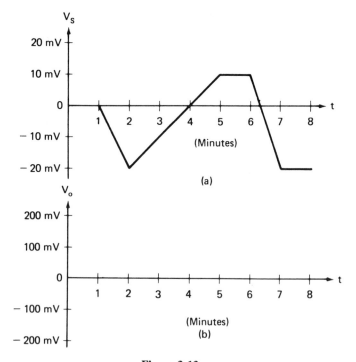

Figure 3-13

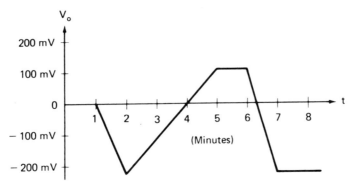

Figure 3-14   Answer to Example 3-4.

Example 3-5

Refer to the circuit described in Example 3-4. If the Op Amp is a 741 C type as listed in Appendix E, what approximate minimum resistance does the signal source V_s see if the impedance to ground from either input is assumed to be infinite?

Answer. As shown in Appendix E, the 741C's minimum A_{VOL} and R_i values are 20,000 and 150 kΩ, respectively. Since the circuit is wired externally for a closed-loop gain of 11, we can find the minimum effective impedance seen by V_s with Eq. (3-7b). Thus in this case

$$R_{i(\text{eff})} \cong \left(\frac{20,000}{11} \right) 150 \text{ k}\Omega \cong 273 \text{ M}\Omega. \qquad (3\text{-}7b)$$

3.4   THE VOLTAGE FOLLOWER

In some applications, an amplifier's voltage gain is not as important as its ability to match a high-internal-resistance signal source to a low, possibly varying, resistance load. The Op Amp in Fig. 3-15 is connected to work as a *voltage follower* which has an extremely large input resistance and is capable of driving a relatively low-resistance load. The voltage follower's output resistance is very small; therefore, variations in its load resistance negligibly affect the amplitude of the output signal. When used between a high-internal-resistance signal source and a smaller, varying-resistance load, the voltage follower is called a *buffer amplifier*.

The voltage follower is simply a noninverting amplifier, similar to the one in Fig. 3-11a, where R_1 is replaced with infinite ohms (an open) and R_F is replaced with zero ohms (a short). (See Fig. 3-16.) Due to the Op Amp's

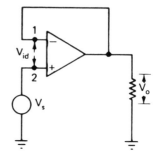

Figure 3-15 Op Amp connected to work as a voltage follower.

large open-loop gain A_{VOL}, the differential input voltage V_{id} is very small; therefore, the inputs 1 and 2 are virtually at the same potential. Since the output signal V_o is the voltage ato input 1, and since the input signal V_s is directly applied to input 2, then

$$V_o \cong V_s.$$

With the input and output signal voltages equal, the voltage gain of the voltage follower is 1 (unity). We can see this another way—by substituting $R_1 = \infty \ \Omega$ into the gain Eq. (3-6). The term R_F/R_1 becomes zero and $A_v = 1$. The term *voltage follower* therefore describes the circuit's function: The output voltage V_o follows the input voltage V_s waveform.

As with the noninverting amplifier, the effective input $R_{in(eff)}$ of the voltage follower is about A_{VOL}/A_v times larger than R_i, where R_i is the differential input resistance measured under open-loop conditions. Since the voltage follower's closed-loop gain $A_v \cong 1$, the ratio A_{VOL}/A_v is very large—equal to A_{VOL}. This accounts for the extremely high input impedance of the voltage follower.

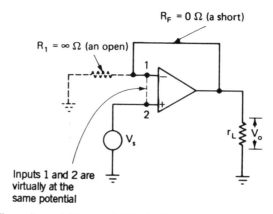

Figure 3-16 The voltage follower of Fig. 3-10 can be shown to be a noninverting amplifier.

As the input impedance $Z_{in(\text{eff})}$ is made more ideal (increased) by the use of negative feedback, the output resistance is also improved (reduced). More specifically, the effective output resistance $R_{o(\text{eff})}$ of an Op Amp is smaller than the output resistance R_o measured with open-loop conditions by the factor A_v/A_{VOL}. That is,

$$R_{o(\text{eff})} \cong \left(\frac{A_v}{A_{VOL}} \right) R_o. \tag{3-8}$$

This shows that the smaller the closed-loop gain A_v, the smaller the effective output resistance $R_{o(\text{eff})}$.

Example 3-6

(a) If the circuit in Fig. 3-15 has the input voltage waveform V_s shown in Fig. 3-8a, sketch the output voltage waveform. Assume that initially the Op amp was nulled.

(b) If a signal source has 100 kΩ of internal resistance and an output of 4 mV *before* being connected to input 2 (open circuited), what is the output of the signal source *after* it is connected to the input 2 of this circuit?

Answers

(a) Since the gain A_v of the voltage follower is unity, and since there is no phase inversion, the output V_o has the same waveform as the input V_s.

(b) The input impedance of the voltage follower is in the range of hundreds of megohms and therefore appears as an open compared to the 100 kΩ internal resistance of the signal source. Thus, there is essentially no signal voltage drop across the internal resistance, and the output of the signal generator remains very close to its 4-mV open-circuit value after being connected to input 2. Therefore the output of this voltage follower is also about 4 mV.

REVIEW QUESTIONS 3

3-1. What two factors outside of the Op Amp affect its maximum unclipped output signal capability?

3-2. When the Op Amp is wired as an inverting amplifier (Fig. 3-5), what is the approximate input resistance as seen by the signal source V_s?

3-3. Why is the Op Amp seldom used as a signal amplifier in open loop?

3-4. If a low-frequency sine wave with a 1-V peak-to-peak amplitude is directly applied to an Op Amp's inputs 1 and 2, what kind of waveform would you expect at the output?

3-5. What is the difference between the closed-loop gain and the open-loop gain?

3-6. How does the differential input signal V_{id} compare in magnitude to the output voltage V_o?

3-7. How can an Op Amp input terminal be virtually grounded but not actually grounded?

3-8. How does the input resistance of a noninverting amplifier compare with the input resistance of the inverting type?

3-9. If an Op Amp is described as having been nulled, what does this mean?

3-10. What effect does negative feedback have on the voltage gain of an amplifier?

3-11. If an Op Amp is connected to have negative feedback, how does its effective output resistance compare with the output resistance specified by the manufacturer?

3-12. If an Op Amp is connected to operate as a noninverting amplifier, with a closed-loop gain lower than the open-loop gain, how does its effective input resistance compare with the input resistance specified by the manufacturer?

3-13. Which would you expect to have larger input impedance: an Op Amp connected to work as a noninverting amplifier with a closed-loop gain of 101, or the same Op Amp connected to work as a voltage follower?

3-14. Since the voltage follower has only unity voltage gain, what good is it?

3-15. What might happen to the output signal voltage waveform if the load resistance is too small?

3-16. What might happen to the output signal voltage waveform if the dc supply voltages are too small?

PROBLEMS 3

Section 3.2

In the circuit of Fig. 3-17a, the signal driving the Op Amp is taken off a bridge circuit as shown. The thermistor in the bridge is a resistance with a high negative temperature coefficient; that is, its resistance significantly

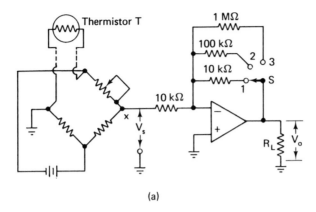

(a)

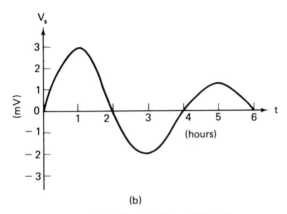

(b)

Figure 3-17   For Problems 3-1–3-4.

decreases or increases as its temperature increases or decreases, respectively. The thermistor is in an oven. As the oven's temperature rises or falls, the voltage V_s swings positively or negatively. The load R_L represents the effective resistance of a motor's armature. When the shaft of the motor turns, it operates a fuel valve and thus admits more or less fuel to the oven. This system therefore controls the temperature in the oven. The switch S provides a means of selecting sensitivity (the closed loop voltage gain).

If the voltage across the bridge, which is voltage V_s, has the waveform shown in Fig. 3-17b, find the voltage gain A_v and select a waveform from Fig. 3-18 that best represents the Op Amp's output V_o for each of the following conditions:

3-1. Switch S is in position 1.

3-2. Switch S is in position 2.

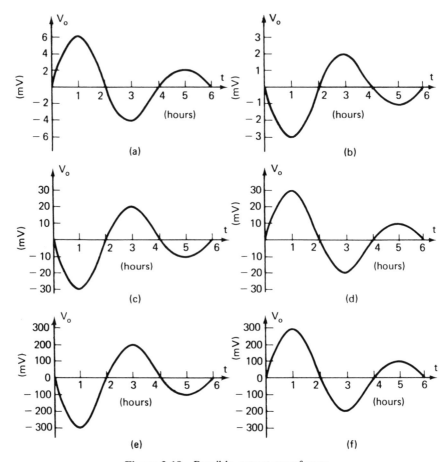

Figure 3-18   Possible output waveforms.

3-3. Switch S is in position 3.

3-4. What is the load resistance across the bridge (points x and ground) in the circuit of Fig. 3-17a?

Section 3.3

If the Op Amp circuit is removed from the system and replaced with the one in Fig. 3-19, but the waveform V_s is still as shown in Fig. 3-17b, find this circuit's voltage gain and select a waveform from Fig. 3-18 that best represents the Op Amp's output V_o for each of the following conditions:

3-5. The switch S is in position I.

3-6. The switch S is in position II.

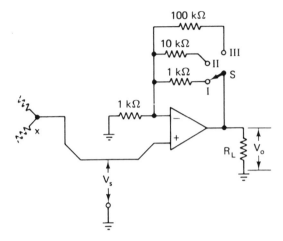

Figure 3-19   For Problems 3-5–3-9.

3-7. The switch S is in position III.

3-8. Which of the Op Amp circuits—the one in Fig. 3-17a or the one in Fig. 3-19—draws the larger current from the bridge?

3-9. Which of the Op Amp circuits—the one in Fig. 3-17a or the one in Fig. 3-19—offers the more constant load on the bridge (keep in mind that the switch S position might frequently be changed)?

The switches S_1 and S_2 in the circuit of Fig. 3-20 are ganged; that is, when S_1 is in position 1, S_2 is also in position 1. Similarly, S_1 and S_2 are in position 2 simultaneously.

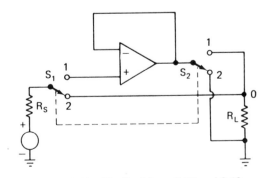

Figure 3-20   For Problems 3-10 and 3-13.

Section 3.4

3-10. If the signal source is 2 V dc, $R_s = 50$ kΩ, and $R_L = 1$ kΩ in the circuit of Fig. 3-20, what is the approximate voltage at point 0 to ground when S_1 is in position 2?

3-11. If the signal source is 3 V dc, $R_s = 150$ kΩ, and $R_L = 1$ kΩ in the circuit of Fig. 3-20, what is the approximate voltage at point 0 with respect to ground when S_1 is in position 2?

3-12. Referring to Problem 3-10, what is the voltage at point 0 with respect to ground when the switches are in position 1?

3-13. Referring to Problem 3-11, what is the voltage at point 0 with respect to ground when the switches are in position 1?

3-14. What is the typical effective output resistance of the 709 Op Amp whose characteristics are given in Appendix J, when it is connected to work as a voltage follower?

3-15. What is the typical effective output resistance of the 741 Op Amp whose characteristics are given in Appendix F, when it is connected to work with a closed-loop gain A_v of 1000?

3-16. Over what range can the gain A_v of the circuit shown in Fig. 3-21a be adjusted if the resistor R can be varied from 0 Ω to 1 MΩ?

Section 3.2

3-17. If $R = 190$ kΩ in the circuit of Fig. 3-21a, what is the output voltage V_o if $V_s = -2$ mV dc? Assume V_o was 0 V when V_s was 0 V.

Section 3.3

3-18. If $V_s = -3$ mV dc and $R = 5$ kΩ in the circuit shown in Fig. 3-21b, what is the output voltage V_o?

3-19. If resistance R can be adjusted from 0 Ω to 200 kΩ in the circuit of Fig. 3-21b, over what range can its voltage gain A_v be varied?

Section 3.2

The following four problems refer to the circuit in Fig. 3-21a with the input voltage waveform as shown in Fig. 3-21c. The Op Amp was initially nulled and has the characteristics in Fig. 3-1.

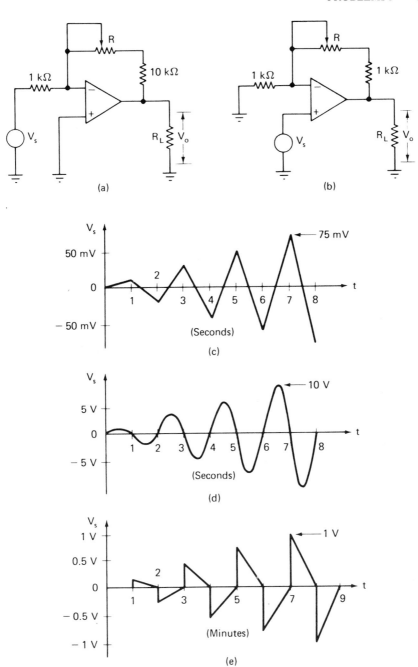

Figure 3-21   Problems 3-16–3-28.

3-20. Sketch the output waveform V_o if the dc supply voltages are $+15$ V and -15 V, the resistance R is adjusted to 90 kΩ, and $R_L = 2$ kΩ.

3-21. Sketch the output waveform V_o if the dc supply voltages are ± 15 V, the resistance R is adjusted to 140 kΩ, and $R_L = 1$ kΩ.

3-22. Referring to Problem 3-20, sketch the output waveform V_o if the dc supply voltages are reduced from ± 15 V to ± 7.5 V. The R and R_L values are unchanged.

3-23. Referring to Problem 3-21, sketch the output waveform V_o if the load resistance R_L is reduced to 200 Ω.

Section 3.3

The following three problems refer to the circuit in Fig. 3-21b with the input voltage as shown in Fig. 3-21d. The Op Amp was initially nulled and has the characteristics in Fig. 3-1.

3-24. Sketch the output waveform V_o if the dc supply voltages are ± 15 V, the resistance R is adjusted to 0 Ω, and $R_L = 2$ kΩ.

3-25. Sketch the output waveform V_o if the dc supply voltages are ± 5 V, the resistance R is adjusted to 0 Ω, and $R_L = 2$ kΩ.

3-26. Sketch the output waveform V_o if the dc supply voltages are ± 15 V, the resistance R is adjusted to 0 Ω, and $R_L = 200$ Ω.

3-27. Referring to Fig. 3-21a, if the circuit has the input waveform shown in Fig. 3-21e, $R = 40$ kΩ, $R_L = 2$ kΩ, ± 15-V supply voltages, and the Op Amp characteristics in Fig. 3-1, show the output voltage waveform.

3-28. Rework the previous problem using the circuit in Fig. 3-21b instead. The values of R, R_L, and supply voltages are unchanged.

4

OFFSET CONSIDERATIONS

The practical Op Amp, unlike the hypothetical ideal version, has some dc output voltage we are calling *output offset* voltage, even though both of its inputs are grounded. Such an output offset is an error voltage and is generally undesirable. The causes and cures of output offset voltages are the subjects of this chapter. Here we will become familiar with parameters that enable us to predict the maximum output offset voltage that a given Op Amp circuit can have. On the foundation laid in this chapter, we will build an understanding of why, and an ability to predict how much, a given Op Amp's output voltage tends to drift with power supply and temperature changes, which are important subjects discussed later.

4.1 INPUT OFFSET VOLTAGE V_{io}

The input offset voltage V_{io} is defined as the amount of voltage required across an Op amp's inputs 1 and 2 to force the output voltage to 0 V. In a previous chapter we learned that ideally, when the differential input voltage $V_{id} = 0$ V, as in Fig. 4-1, the output offset V_{oo} is 0 V too. In the practical case, however, V_{oo} voltage will always be present. This is caused by imbalances within the Op amp's circuitry. They occur because the transistors of the input differential stage within the Op Amp usually admit slightly different collector currents even though both bases are at the same potential. This causes a differential output voltage from the first stage, which is amplified and possibly aggravated by more imbalances in the following stages. The combined result of these imbalances is the output offset voltage V_{oo}. If small dc voltage of proper polarity is applied to inputs 1 and 2, it will decrease the

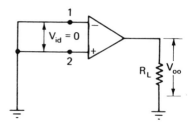

Figure 4-1 Ideally, $V_{oo} = 0$.

output voltage. The amount of differential input voltage, of the correct polarity, required to reduce the output to 0 V is the input offset voltage V_{io}. When the output is forced to 0 V by the proper amount and polarity of input voltage, the circuit is said to be *nulled* or *balanced*. The polarity of the required input offset voltage V_{io} at input 1 with respect to input 2 might be positive as often as negative. This means that the output offset voltage, before nulling, can be positive or negative with respect to ground.

A typical nulling circuit of inverting-mode amplifiers is shown in Fig. 4-2. An equivalent for noninverting amplifiers is shown in Fig. 4-3. In each of these circuits, the potentiometer POT can be adjusted to provide the value

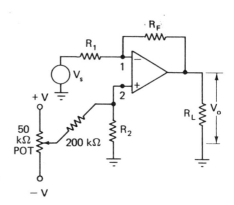

Figure 4-2 Typical null balancing circuit in inverting amplifiers.

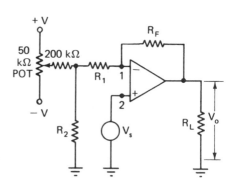

Figure 4-3 Typical null balancing circuit in noninverting amplifiers.

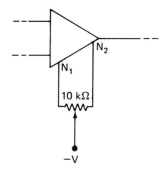

Figure 4-4   Available null adjustment for a 741 Op Amp.

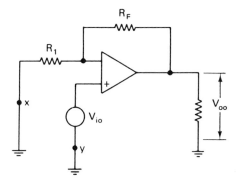

Figure 4-5   Equivalent circuit of either the inverting or the noninverting amplifier.

and polarity of dc input voltage required to null the output to 0 V. As shown, the ends of the POT are connected to the positive and negative rails. With some Op Amps, such as types 748, 777, 201, 741,* etc. (see Appendix E), *offset adjust* pins are provided. A POT can be placed across the OFFSET NULL pins of a 741, as shown in Fig. 4-4. Adjustment of this POT will null the output if the closed-loop gain A_v is not too large.

If an Op Amp circuit is not nulled, more or less output offset voltage V_{oo} exists, depending on its specified input offset voltage V_{io}, closed-loop gain A_v, and other factors that are discussed in the following sections.

To predict how much output offset V_{oo} a given Op Amp circuit will have, "caused" by its input offset voltage V_{io}, a noninverting amplifier model can be used—see Fig. 4-5. In other words, if $V_B = 0$ V with either the inverting or noninverting amplifier circuit, its model is the circuit of Fig. 4-5. As shown in this figure, the Op Amp's specified input offset voltage V_{io} is equivalent to a dc signal source working into a noninverting type amplifier. If this model in

*Generally, 741-type Op Amps have characteristics such as listed in Appendix F, 741s are identified in several ways, depending on the manufacturer. To name a few: Fairchild μA741, Motorola MC1741, National Semiconductor LM741, Raytheon RM741, Signetics μA741.

Fig. 4-5 represents an inverting amplifier, the signal source V_s is replaced with its own internal resistance at point x. If this equivalent circuit represents a noninverting amplifier, V_s is replaced with its own internal resistance at point y. A short circuit replaces V_s if its internal resistance is negligible. According to Fig. 4-5 then, we can show that the output offset voltage V_{oo}, "caused by" the input offset voltage, is the product of the closed-loop gain and the specified V_{io}:

$$\boxed{V_{oo} = A_v V_{io},} \qquad (4\text{-}1)$$

where

$$A_v \cong \frac{R_F}{R_1} + 1. \qquad (3\text{-}6)$$

Example 4-1

If the Op Amp in the circuit of Fig. 4-6 is a 741 type (see appendix F), what is the *maximum* possible output offset voltage V_{oo}, caused by the input offset voltage V_{io} before any attempt is made to null the circuit with the 10-kΩ POT?

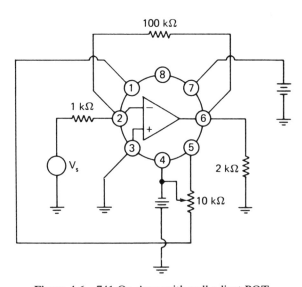

Figure 4-6 741 Op Amp with null adjust POT.

Answer. According to the specifications of the 741, its maximum $V_{io} = 6\text{ mV}$. Since the input resistance $R_1 = 1\text{ k}\Omega$, assuming that the internal resistance of V_s is negligible and since the feedback resistance $R_F = 100\text{ k}\Omega$, the input offset voltage is multiplied by

$$A_v \cong \frac{R_F}{R_1} + 1 = \frac{100\text{ k}\Omega}{1\text{ k}\Omega} + 1 = 101.$$

Therefore, before the circuit is nulled, we might have an output offset as large as

$$V_{oo} = A_v V_{io} \cong 101(6\text{ mV}) = 606\text{ mV, or about } 0.6\text{ V}.$$

This means that the output voltage with respect to ground can be either a positive or a negative 0.6 V even though the input signal $V_s = 0$ V.

4.2   INPUT BIAS CURRENT I_B

Most types of IC Op Amps have two transistors in the first (input) differential stage. Transistors, being current-operated devices, require some base bias currents. Therefore, small dc bias currents flow in the input leads of the typical Op Amp as shown in Fig. 4-7. An input bias current I_B is usually specified on the Op Amp specification sheets as shown in the Appendices. It is defined as the average of the two base bias currents; that is,

$$I_B = \frac{I_{B_1} + I_{B_2}}{2}. \tag{4-2}$$

These base bias currents, I_{B_1} and I_{B_2} are *about* equal to each other, and therefore the specified input bias current I_B is *about* equal to either one of them; that is,

$$I_B \cong I_{B_1} \cong I_{B_2}. \tag{4-3}$$

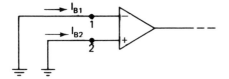

Figure 4-7 Base bias currents flow into the Op Amp and return to ground through the power supplies.

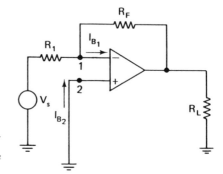

Figure 4-8 Current I_{B_1} sees some resistance on its way to ground, whereas I_{B_2} sees a short.

Depending on the type of Op Amp, the value of input bias current is usually small, generally in the range from a few to a few hundred nanoamperes in general-purpose, economical Op Amps. Though this seems insignificant, it can be a problem in circuits using relatively large feedback resistors, as we will see.

A difficulty that the base bias currents might cause can be seen if we analyze their effect on a typical amplifier such as in Fig. 4-8. Note that I_{B_2} flows out of ground directly into input 2. (See Fig. 1-8 for the current paths of a typical differential input stage.) Therefore, input 2 is 0 V with respect to ground. On the other hand, base bias current I_{B_1} sees resistance on its way to input 1, that is, the parallel paths containing R_1 and R_F have a total effective resistance as seen by the base bias current I_{B_1}. The flow of I_{B_1} through this effective resistance causes a dc voltage to appear at input 1. With a dc voltage to ground at input 1 while input 2 is 0 V to ground, a dc differential input voltage V_{id} appears *across* these inputs. This dc input amplified by the closed-loop gain causes an output offset voltage V_{oo}. The amount of output offset voltage V_{oo}, caused by the base bias current I_{B_1}, can be approximated with the equation

$$V_{oo} \cong R_F I_B,$$

(4-4)*

where $I_B \cong I_{B_1}$, and is usually specified by the manufacturer.

Example 4-2

Referring to the circuit in Fig. 4-6, what maximum output offset voltage might it have, caused by the base bias current, before it is nulled? The Op Amp is a type 741.

---

*See Appendix G for the derivation.

Answer. According to the 741's specification sheets (Appendix F), the maximum input bias current $I_B = 500$ nA. Since $R_F = 100$ kΩ, the base bias current could cause an output offset as large as

$$V_{oo} \cong R_F I_B = 100 \text{ k}\Omega(500 \text{ nA}) = 50 \text{ mV}. \qquad (4\text{-}4)$$

We should note in these last two examples that the output offset caused by the input offset voltage V_{io} is about 10 times larger than the output offset caused by the input bias current I_B. In this case then, the input offset voltage V_{io} is potentially the greater problem. However, according to Eq. (4-4), we can see that the output offset V_{oo} caused by the bias current I_B is larger, and therefore potentially more troublesome, with larger values of feedback resistance R_F.

Example 4-3

Referring again to the circuit of Fig. 4-6, suppose that we replace the 1-kΩ input resistor with 100 kΩ and the 100-kΩ feedback resistor with 10 MΩ:

(a) How does this affect the closed-loop gain A_v of this circuit compared to what it was originally? Before any attempt is made to null this circuit, find

(b) Its maximum output offset caused by the input offset voltage V_{io}, and

(c) The maximum output offset caused by the input bias current I_B. The Op Amp is still a type 741.

Answers

(a) The closed-loop gain A_v is unchanged; that is, this circuit's gain is very nearly equal to the ratio R_F/R_1, which was not changed.

(b) With no change in the closed-loop gain, the output offset caused by the input offset voltage does not change [see Eq. (4-1)].

(c) The output offset caused by input bias current I_B is larger with circuits using larger feedback resistors. according to Eq. (4-4). Thus in this case

$$V_{oo} \cong R_F I_B = 10 \text{ M}\Omega(500 \text{ nA}) = 5 \text{ V}.$$

This is a relatively large output offset. And approaches the positive rail of some Op Amp circuits. This example presents a good case for the use of small feedback resistors.

The effect of the input bias current I_B on the output offset voltage can be minimized if a resistor R_2 is added in series with the noninverting input as

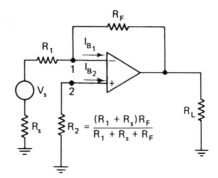

Figure 4-9 Resistor R_2 in series with the noninverting input reduces the output offset voltage caused by the input bias current I_B.

shown in Fig. 4-9. By selecting the proper value of R_2, we can make the resistance seen by the base bias current I_{B_2} equal to the resistance seen by the base bias current I_{B_1}. This will raise input 2 to the dc voltage at input 1. In other words, if the currents I_{B_1} and I_{B_2} are equal, and the resistances seen by these currents are equal, the voltages to ground at inputs 1 and 2 are equal. This means that there will be no dc differential input voltage V_{id} to cause an output offset V_{oo}. The value of R_2 needed to eliminate or reduce the dc differential input voltage V_{id} and the resulting output offset V_{oo} is easily found.

Note in Fig. 4-9 that current I_{B_1} sees two parallel paths, one containing R_1 and R_s in series and the other containing R_F. Thus when $V_o \cong 0\ V$, current I_{B_1} sees a resistance $R_1 + R_s$ in parallel with the feedback resistor R_F. Since current I_{B_2} is to see a resistance R_2 that is equal to the total resistance seen by I_{B_1}, we can show that

$$R_2 = \frac{(R_1 + R_s)R_F}{R_1 + R_s + R_F}. \tag{4-5a}$$

If the internal resistance of the signal source is zero, then

$$R_2 = \frac{R_1 R_F}{R_1 + R_F}. \tag{4-5b}$$

Example 4-4

Referring to the circuit in Fig. 4-6, what value of resistance can we use in series with the noninverting input (pin 3) to reduce or eliminate the output offset voltage caused by the input bias current? Assume that the internal resistance of the signal source V_s is negligible.

Answer. Since the input resistor $R_1 = 1$ kΩ, the feedback resistor $R_F = 100$ kΩ, and the signal source's resistance $R_s = 0$ Ω, we can use

$$R_2 = \frac{(1 \text{ k}\Omega + 0)100 \text{ k}\Omega}{1 \text{ k}\Omega + 0 + 100 \text{ k}\Omega} = 990 \text{ }\Omega.$$

4.3 INPUT OFFSET CURRENT I_{io}

In the previous section we learned that a resistor R_2 placed in series with the noninverting input 2 reduces the output offset caused by the input bias current. However, Eq. (4-5) for finding the necessary value of R_2 was derived assuming that the base bias currents I_{B_1} and I_{B_2} are equal. In practice, due to imbalances within the Op Amp's circuitry, these currents are at best only approximately equal, as indicated in Eq. (4-3). The input offset current I_{io}, usually specified by the manufacturer, is a parameter that indicates how far from being equal the currents I_{B_1} and I_{B_2} can be. In fact, the input offset current is defined as the difference in the two base bias currents; that is,

$$I_{io} = |I_{B_1} - I_{B_2}|. \qquad (4\text{-}6)$$

When given the value of input offset current I_{io}, we can predict how much output offset voltage a circuit like that in Fig. 4-9 might have, caused by the existence of base bias currents. Due to inequalities of the currents I_{B_1} and I_{B_2}, the voltages with respect to ground at inputs 1 and 2 will be unequal, even though a properly chosen value of R_2 is used. This causes a dc differential input voltage V_{id}, which in turn causes an output offset voltage V_{oo}. In other words, the amount of dc differential input voltage and the amount of resulting output offset depend on the amount of difference in the base bias currents I_{B_1} and I_{B_2}, which is the input offset current I_{io}. More specifically, the amount of output offset voltage V_{oo} in a circuit like that in Fig. 4-9, caused by the input offset current I_{io}, can be closely approximated with the equation

$$V_{oo} \cong R_F I_{io} \qquad (4\text{-}7)^*$$

if the value of R_2 is determined with Eq. (4-5).

* See Appendix H for the derivation

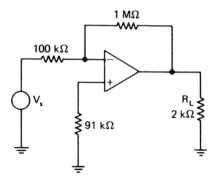

Figure 4-10   Circuit for Example 4-5.

Example 4-5

Referring to the circuit in Fig. 4-10, find the maximum output offset caused by the input offset current I_{io}. The Op Amp is a type 741, and the internal resistance of the signal source V_s is negligible.

Answer. The 741's maximum input offset current I_{io} = 200 nA as indicated on its specification sheet. Since the value of R_2 satisfies Eq. (4-5), we can closely estimate the output offset caused by I_{io} with Eq. (4-7). In this case

$$V_{oo} \cong R_F I_{io} = 1\ M\Omega(200\ nA) = 200\ mV.$$

This shows that we might have an output offset due to unequal base bias currents even though a properly chosen resistor R_2 is placed between the noninverting input and ground. If R_2 is not used, however, the output offset caused by input currents is considerably larger and therefore potentially a greater problem.

Example 4-6

Referring again to the circuit in Fig. 4-10, find the maximum output offset caused by the input currents if R_2 is removed and, instead, the noninverting input is connected directly to ground. As before, the Op Amp is a type 741.

Answers

$$V_{oo} \cong R_F I_B = 1\ M\Omega(500\ nA) = 500\ mV.$$

It is interesting to note that the output offset can be more than doubled if the resistor R_2 is not used. R_2 may or may not be necessary, depending on how important it is to have little or no output offset. The required stability of the output voltage with temperature changes must be considered too, as we will see in a later chapter. Some types of IC Op Amps are made with FETs or high-beta transistors* in the input differential stage. These have dc input currents on the order of just a few nanoamperes. Some *hybrid* models of Op Amps, which contain discrete and integrated circuits, have input bias currents in the order of 0.01 pA or less. Of course, such a small dc input bias current reduces the current-generated output offset voltage to insignificance. As we will see in applications later, an extremely small input bias current is desirable, and in fact necessary, in long-term integrating and sample-and-hold circuits.

4.4 COMBINED EFFECTS OF V_{io} AND I_{io}

In a circuit such as that in Fig. 4-9, the input offset voltage V_{io} can cause either a positive or a negative output offset voltage V_{oo} [see Eq. (4-1)]. Likewise, the input offset current I_{io} can cause either a positive or a negative output offset voltage V_{oo} [see Eq. (4-7)]. The effects of input offset voltage V_{io} and input offset current I_{io} might buck and cancel each other, resulting in little output offset. On the other hand, they might be additive, causing an output offset voltage V_{oo} that is the sum of the output offsets caused by V_{io} and I_{io} working independently. Thus the total output offset voltage in the circuit of Fig. 4-9 can be as large as

$$V_{oo} \cong A_v V_{io} + R_F I_{io} \qquad (4\text{-}8a)$$

if the value of R_2 satisfies Eq. (4-5). This equation is sometimes shown in other equivalent forms, such as

$$V_{oo} \cong [V_{io} + R_2 I_{io}]\left(\frac{R_F}{R_1} + 1\right) \qquad (4\text{-}8b)$$

or

$$V_{oo} \cong [V_{io} + R_2(I_{B_1} - I_{B_2})]\left(\frac{R_F}{R_1} + 1\right) \qquad (4\text{-}8c)$$

---

*Op Amps with high-beta input transistors are sometimes called *super beta* transistor input Op Amps. An extremely high beta can be achieved by connecting conventional transistors in a Darlington pair.

Example 4-7

Considering the effects of both V_{io} and I_{io}, what is the maximum output offset voltage V_{oo} of the circuit in Fig. 4-10 if the Op Amp is a type 741?

Answer. Arbitrarily selecting Eq. (4-8b), we can show that the maximum output offset voltage is

$$V_{oo} \cong [6 \text{ mV} + 91 \text{ k}\Omega(200 \text{ nA})]\left(\frac{1000 + 100}{100}\right) = 266 \text{ mV}.$$

REVIEW QUESTIONS 4

4-1. What is the input offset voltage of an Op Amp?

4-2. Why is an input offset voltage V_{io} often needed to null the output of an Op Amp?

4-3. What polarity of required input offset voltage would you expect at input 1 with respect to input 2?

4-4. What is the meaning of the term *base bias current*?

4-5. What are the difference and similarity of the base bias and input bias currents?

4-6. What problem might exist if a significant base bias current exists and if the feedback resistor is relatively large?

4-7. How can the undesirable effect of a significant input bias current be minimized?

4-8. The nulling circuits in Figs. 4-2 and 4-3 are able to provide a dc differential input voltage that is (*positive*), (*negative*), (*either polarity*) at input 1 with respect to input 2.

4-9. The nulling circuits in Figs. 4-2 and 4-3 are able to adjust the output V_o to zero volts in spite of the existence of (*input offset voltage*), (*input bias current*), (*input offset voltage and input bias current*).

4-10. What is the meaning of the term *input offset current*?

PROBLEMS 4

4-1. If the Op Amp in the circuit of Fig. 4-11 is a type 741, what is the circuit's closed-loop voltage gain if the potentiometer (POT) is adjusted to 39 kΩ? The eight-pin DIP (dual-in-line) package has the same pin identifications as the eight-pin circular metal can package.

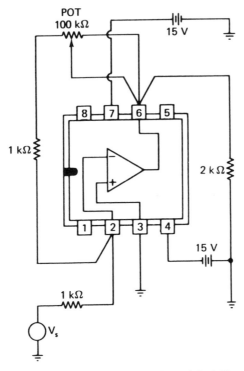

Figure 4-11   Circuit for Problems 4-1–4-13.

4-2. If the Op Amp in the circuit of Fig. 4-11 is a LF 13741 type (Appendix E), what is the circuit's closed-loop gain if the POT is adjusted to 14 kΩ? It has the same pin identifications as does the 741 eight-pin circular metal can package or DIP.

4-3. If the slide on the POT is moved to the far right in the circuit of Fig. 4-11, what is the approximate closed-loop voltage gain of the circuit?

4-4. If the slide on the POT is moved to the far left in the circuit of Fig. 4-11, what is the approximate closed-loop voltage gain of the circuit?

Section 4.1

4-5. In the circuit described in Problem 4-1, what is the maximum output offset caused by the Op Amp's input offset voltage?

4-6. What is the maximum output offset caused by the input offset voltage in the circuit described in Problem 4-2?

Section 4.2

4-7. In Problem 4-1, what is the circuit's maximum output offset caused by the existence of base bias currents in the input leads?

4-8. What is the maximum possible output offset voltage caused by the existence of input bias currents in the circuit described in Problem 4-2?

Section 4.4

4-9. Referring to the circuit described in Problem 4-1, what maximum output offset voltage might it have with the combined effects of input offset voltage and input bias currents?

4-10. Referring to the circuit described in Problem 4-2, what maximum output offset voltage might it have with the combined effects of input offset voltage and input bias currents?

Section 4.3

4-11. If the POT in the circuit of Fig. 4-11 is replaced with a fixed 100-kΩ resistor, what value of resistance should we put in series with pin 3 and ground to reduce the effect of input bias currents? Assume that the Op Amp is a type 741 and that the internal resistance of the signal source is negligible.

4-12. If the POT in the circuit of Fig. 4-11 is replaced with a fixed 10-kΩ resistor, what value of resistance should we put in series with pin 3 and ground to reduce the effect of input bias current? The Op Amp is a type LF 13741 and the internal resistance of the signal source is negligible.

4-13. In the circuit described in Problem 4-11, what is the maximum output offset voltage caused by the existence of input bias currents if the properly selected resistance is used between pin 3 and ground?

4-14. In the circuit described in Problem 4-12, what is the maximum output offset caused by the existence of input bias currents if the properly selected resistance is used between pin 3 and ground?

Review

4-15. In the circuit of Fig. 4-12, if full-range adjustment of the 50-kΩ POT almost, but not quite, nulls the output, what small modification can we make to have full null control?

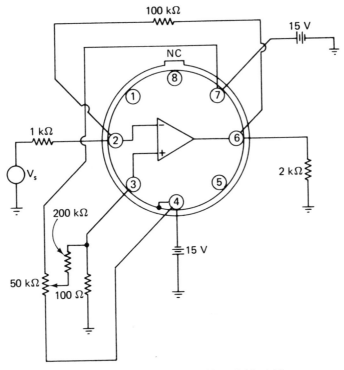

Figure 4-12 Circuit for Problem 4-14–4-18.

4-16. If in the circuit of Fig. 4-12 the input signal V_s is a sine wave but the output has the waveform shown in Fig. 4-13, what modification or adjustment will probably correct the problem?

4-17. If the circuit in Fig. 4-12 has a maximum peak-to-peak output capability of 20 V, what maximum peak-to-peak input signal V_s can we apply?

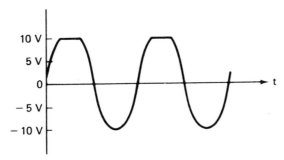

Figure 4-13 Wave form for Problems 4-16 and 4-20.

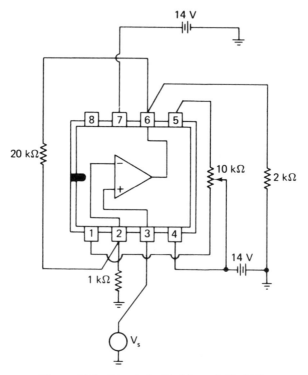

Figure 4-14   Circuit for Problems 4-19–4-22.

4-18. If the circuit in Fig. 4-12 is nulled, has a maximum peak-to-peak output capability of 20 V, and has an input sine-wave voltage V_s of 2 V peak to peak, what modification is necessary in this circuit if clipping of the output signal is to be avoided?

4-19. If V_s is a sine wave in the circuit of Fig. 4-14, and if an oscilloscope reading shows a sine-wave voltage at pin 6 to ground with a peak-to-peak value of 6.3 V, what is the peak-to-peak amplitude of the input V_s?

4-20. If the circuit in Fig. 4-14 has the waveform shown in Fig. 4-13 at pin 6 with respect to ground, what is the approximate peak-to-peak amplitude of V_s?

4-21. What is the purpose of the 10-kΩ POT in the circuit of Fig. 4-14 if the Op Amp is a type 741?

4-22. If the Op Amp in the circuit of Fig. 4-14 is a type 741, what is its maximum peak-to-peak output signal capability?

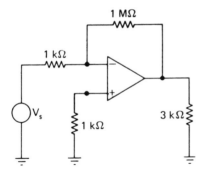

Figure 4-15 Circuit for Problems 4-23 and 4-24.

Section 4.4

4-23. If the Op Amp in the circuit of Fig. 4-15 is a type 741, what is its maximum possible output offset, caused by both the input offset voltage and the input offset current?

4-24. If the Op Amp in the circuit of Fig. 4-15 is a type LF 13741, what is its maximum possible output offset, caused by both the input offset voltage and the input offset current?

5

COMMON-MODE VOLTAGES
AND DIFFERENTIAL AMPLIFIERS

Industrial and commercial environments are electrically noisy. Lighting fixtures, motors, relays, and switches, to name just a few, emit time-varying electric and magnetic fields. These fields induce electrical noise into nearby conductors. Noise voltages (V_n) induced into conductors at inputs of amplifiers are amplified along with the signals (V_s) that are intended to be amplified. Thus the amplitudes of both the induced noise V_n and signal V_s are increased together. In many amplifier applications, the leads and interconnections can be kept short and the amplitude of the induced noise, relative to the signal, is small and of no consequence. In applications that require that signals be transported over distances, the induced noise can be very troublesome. For example, remote sensors monitoring temperatures, pressures, tensile stress, etc. are commonly used in industrial and medical equipment. In such applications the induced noise voltages are often larger than the signals that we are trying to amplify. By use of amplifiers with differential inputs, we can make induced noise voltages *common mode*, and as you will see, this will greatly reduce their amplitudes compared to the signals.

A voltage at both inputs of an Op Amp, as shown in Fig. 5-1, is a *common-mode* voltage V_{cm}. A common-mode voltage V_{cm} can be dc, ac, or a combination (superposition) of dc and ac. When Op Amps work in time-varying magnetic and electric fields, noise voltages are induced into both input leads. In such cases, dc and ac common-mode voltages V_{cm} can exist simultaneously at an Op Amp's inputs. The induced ac voltage includes 60 Hz from nearby electrical power equipment and include higher frequencies or even irregular transients. In any case, it is undesired and considered as *noise* in the system.

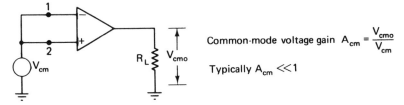

Common-mode voltage gain $A_{cm} = \dfrac{V_{cmo}}{V_{cm}}$

Typically $A_{cm} \ll 1$

Figure 5-1   Op Amp with common-mode voltage applied.

Amplifiers with differential inputs (e.g., Op Amps) have more or less ability to reject common-mode voltages. This means that, although fairly large common-mode voltages might exist at an Op Amp's differential inputs, these voltages can be reduced to very small and often insignificant amplitudes at the output. In this chapter we will become familiar with manufacturers' specifications that will enable us to predict how well a given Op Amp will reject (not pass) common-mode voltages. We will also see applications in which the Op Amp's ability to reject common-mode voltages is useful.

5.1 DIFFERENTIAL-MODE OP AMP CIRCUIT

To appreciate fully how a properly wired Op Amp circuit is able to reject undesirable common-mode voltages, we should first look at the familiar circuits that have induced noise at their inputs. For example, for both circuits shown in Fig. 5-2, voltage V_s is the desired signal to be amplified. If the input lead has any appreciable length, and if varying fields are present, noise voltage V_n will be induced into it as shown. In both of these circuits, the Op Amp cannot distinguish noise V_n from desired signal V_s; both are amplified and appear at the outputs. Thus if the closed-loop gain A_v of each of these circuits is 100, the output noise voltage V_{no} and output signal voltage V_o are 100 times larger than their respective inputs.

Noise voltages at the output of an Op Amp are greatly reduced if the circuit is connected to operate in a differential mode, as shown in Fig. 5-3a In this circuit, the desired signal V_s is amplified normally because it is applied *across* the two inputs. That is, signal V_s causes a differential input V_{id} to appear across the input terminals 1 and 2. The noise voltage V_n, however, is induced into each input lead with respect to ground or common. With proper component selection, both of the induced noise voltages V_n are equal in amplitude and phase and therefore are common-mode voltages, as shown in Fig. 5-3b. Thus if the noise voltage to ground at input 1 is the same as the noise voltage to ground at input 2, the differential noise voltage across input terminals 1 and 2 is zero. Since ideally an Op Amp amplifies only differential input voltages no noise voltage should appear at the output. However, due to

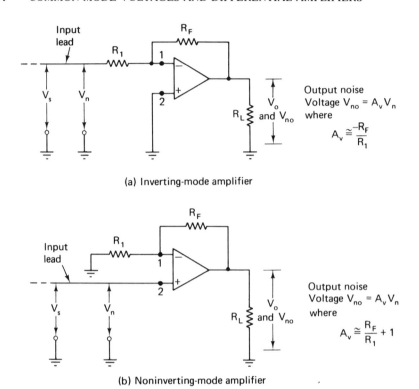

(a) Inverting-mode amplifier

(b) Noninverting-mode amplifier

Figure 5-2   Inverting-mode and noninverting-mode circuits amplify induced noise.

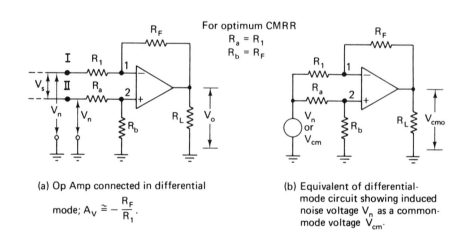

(a) Op Amp connected in differential mode; $A_V \cong -\dfrac{R_F}{R_1}$.

(b) Equivalent of differential-mode circuit showing induced noise voltage V_n as a common-mode voltage V_{cm}.

Figure 5-3   (a) Op Amp connected in differential mode; $A_V \cong -R_F/R_1$. (b) Equivalent of differential-mode circuit showing induced noise voltage V_n as a common-mode voltage V_{cm}.

inherent electrical characteristics of an Op Amp—such as internal capacitances—some common-mode signals will get through to the load R_L. The ratio of the output common-mode voltage V_{cmo} to the input common-mode voltage V_{cm} is called the common-mode gain A_{cm}; that is,

$$A_{cm} = \frac{V_{cmo}}{V_{cm}}. \tag{5-1}$$

Ideally, this gain A_{cm} is zero. In practice, it is finite but usually much smaller than 1.

5.2 COMMON-MODE REJECTION RATIO, CMRR

Op Amp manufacturers do not list a common-mode gain factor. Instead, they list a *common-mode rejection ratio*, CMRR. The CMRR is defined in several essentially equivalent ways by the various manufacturers. It can be defined as the ratio of the change in the input common-mode voltage ΔV_{cm} to the resulting change in input offset voltage ΔV_{io}. Thus

$$CMRR = \frac{\Delta V_{cm}}{\Delta V_{io}}. \tag{5-2}$$

It also is shown to be equal to the ratio of the closed-loop gain A_v to the common-mode gain A_{cm}; that is,

$$CMRR = \frac{A_v}{A_{cm}}. \tag{5-3}$$

Generally, larger values of CMRR mean better rejection of common-mode signals and are therefore more desirable in applications where induced noise is a problem. Later, we will see that the CMRR of an Op Amp tends to decrease with higher common-mode frequencies.

The common-mode rejection is usually specified in decibels (dB), where

$$CMR \ (dB) = 20 \log \frac{\Delta V_{cm}}{\Delta V_{io}} \tag{5-4a}$$

or

$$\text{CMR (dB)} = 20 \log \frac{A_v}{A_{cm}} = 20 \log \text{CMRR}. \qquad (5\text{-}4b)$$

A chart for converting common-mode rejection from a ratio to decibels, or vice versa, is given in Fig. 5-4.

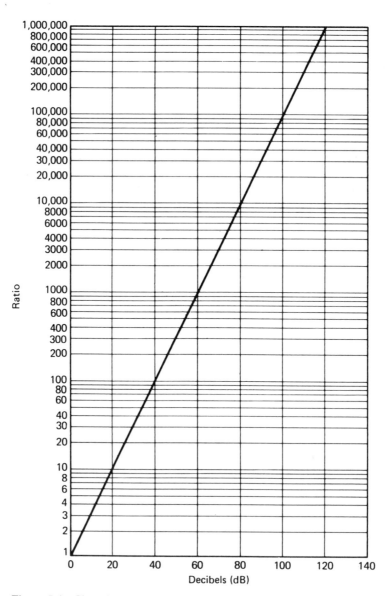

Figure 5-4 Chart for converting voltages ratios to decibels and vice versa.

Equation (5-3) shows that the common-mode voltage gain A_{cm} of a differential-mode Op Amp circuit is a function of its closed-loop gain and the Op Amp's specified CMRR. Rearranging Eq. (5-3), we can show that

$$A_{cm} = \frac{A_v}{\text{CMRR}}. \tag{5-3}$$

The equivalent of this in decibels can be shown as

$$\boxed{20 \log A_{cm} = 20 \log A_v - 20 \log \text{CMRR}} \tag{5-5a}$$

or as

$$\boxed{A_{cm} \, (\text{dB}) = A_v \, (\text{dB}) - \text{CMR} \, (\text{dB}).} \tag{5-5b}$$

Since the closed-loop gain A_v is usually much smaller than the CMRR, the common-mode gain A_{cm} in Eq. (5-3) is much smaller than 1. Therefore, in either version of Eq. (5-5) then, the common-mode gain in decibels is negative.

Some manufacturers define CMRR as the reciprocal of the right side of Eq. (5-2). This implies that the CMRR in decibels is a negative value. If we assume that the specified value of CMR (dB) is positive, then we subtract it from A_v (dB) to obtain the value of A_{cm} (dB), as shown in Eq. (5-5b). If we assume that CMR (dB) is negative, then we add it to A_v (dB) to get A_{cm} (dB). In either case, we solve for the *difference* in the values on the right side of Eq. (5-5) to determine A_{cm} (dB).

The signal voltage gain V_o/V_s of the differential circuit in Fig. 5-3a is determined with Eq. (3-4); that is,

$$|A_v| = \frac{V_o}{V_s} \cong \frac{R_F}{R_1}. \tag{3-4}$$

Whether this gain is inverting or noninverting is not predictable when calculating the common-mode gain A_{cm}. From the signal's point of view, however, this gain is inverting if we reference signal V_s at input I with respect to input II. If V_s is referenced at input II with respect to input I, this gain A_v is noninverting.

Note in Fig. 5-3a that the resistors R_a and R_b are equal to R_1 and R_F, respectively. Also, the resistances to ground looking back toward the source of V_s from inputs I and II must be equal, and likewise the lead lengths are equal. These design criteria establish conditions that make the noise voltages at these two inputs as equal as possible, that is, common mode. The resistors R_a and R_b also serve the same function as does R_2 in Fig. 4-9. These reduce the output offset V_{oo} caused by bias current I_B.

Example 5-1

Referring to the circuit in Fig. 5-3a, if $R_1 = R_a = 1 \text{ k}\Omega$, $R_F = R_b = 10 \text{ k}\Omega$, $V_s = 10$ mV at 1000 Hz, and $V_n = 10$ mV at 60 Hz, what are the amplitudes of the 1000-Hz signal and the 60-Hz noise at the load R_L? The Op Amp's CMR (dB) = 80 dB.

Answer. This circuit's closed-loop gain is

$$|A_v| \cong \frac{R_F}{R_1} = \frac{10 \text{ k}\Omega}{1 \text{ k}\Omega} = 10.$$

Since the 1000 Hz is applied in a differential mode, it is amplified by this gain factor. Thus, at 1000 Hz, the output signal is

$$V_o \cong 10(10 \text{ mV}) = 100 \text{ mV}.$$

The 60-Hz noise voltage, on the other hand, is applied in common mode. Therefore, we first find the common-mode gain with Eq. (5-3); that is,

$$A_{cm} = \frac{A_v}{\text{CMRR}} \cong \frac{10}{10,000} = 1 \times 10^{-3},$$

where 80 dB is equal to the ratio 10,000.
    We can find the amount of common-mode output voltage with Eq. (5-1); that is,

$$V_{cmo} = A_{cm}V_{cm} \cong 1 \times 10^{-3}(10 \text{ mV}) = 10 \text{ } \mu\text{V}.$$

Thus, the 60-Hz output is smaller than the input induced 60 Hz, by the factor 1000, and the Op Amp's differential input is effective in reducing noise problems. V_{cmo} can be in phase with V_{cm} as often as out of phase.
    We could have determined the common-mode output V_{cmo} by using Eq. (5-5). For example, since a closed-loop gain of 10 is equivalent to 20 dB, then by Eq. (5-5b),

$$A_{cm} \text{ (dB)} = 20 - 80 = -60 \text{ dB}.$$

The negative sign means that the common-mode voltage is attenuated by 60 dB; that is, V_{cmo} is smaller than V_{cm} by the factor 1000.

5.3 MAXIMUM COMMON-MODE INPUT VOLTAGES

A common-mode input voltage can be dc as in the circuit of Fig. 5-5. In this case, the bridge circuit has one dc supply voltage E. By the voltage divider action of resistances R, a portion of this supply E is applied to inputs I and II. The dc voltages at these inputs are shown as V_I and V_{II}. Their average value is the dc common-mode in put; that is,

$$V_{cm} = \frac{V_I + V_{II}}{2}. \tag{5-6}$$

Further voltage-divider action takes place by resistors R_a and R_b and by resistors R_1 and R_F, and the actual voltage applied to the noninverting input 2 is

$$V_2 = \frac{V_{II} R_b}{R_a + R_b}. \tag{5-7}$$

Since V_1 is *virtually* at th same potential as V_2, then we can similarly show

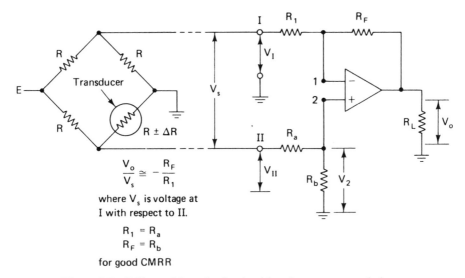

Figure 5-5 Differential-mode circuit with a dc common-mode input.

that

$$V_1 \cong \frac{V_{II} R_b}{R_a + R_b}.$$

(5-8)

The voltage V_1 or V_2 should never exceed the maximum *input voltage* or the maximum *common-mode voltage* specified on an Op Amp's data sheets.

5.4   OP AMP INSTRUMENTATION CIRCUITS

Frequently, in instrumentation and industrial applications, the Op Amp is used to amplify signal output voltages from bridge circuits. We considered such a circuit earlier in Fig. 3-17a. Similar circuits are the useful *half-bridge* circuits as shown in Fig. 5-6. Because the inputs of these circuits are referenced to ground, they are prone to induced noise problems. Therefore, if the bridge-type transducer circuit is required to work in time-varying electric and magnetic fields, its amplifier should be a differential type, such as in Fig. 5-5.

In the circuit of Fig. 5-5, the maximum specified common-mode input voltage of the Op Amp dictates the maximum allowable dc source voltage E on the bridge. The transducer's resistance changes by ΔR when the appropriate change in its physical environment occurs. The bridge converts this physical change to a voltage change across points I and II which is amplified by the Op Amp. Thus, depending on the type of transducer used, a change in temperature, light intensity, strain, or whatever, is converted to an amplified electrical signal. The gain of the differential amplifier can be increased or decreased if we increase or decrease *both* R_F and R_b by equal amounts. In other words, R_F and R_b must remain equal if good *CMR* is to be retained. Gain control by varying *two* resistances simultaneously is not practical for a number of reasons.. If R_F and R_b are changed, the resistances looking into points I and II will change noticeably too. The accompanying problems with this kind of variable loading on the bridge can be avoided by use of high input impedance circuits such as in Fig. 5-7. The gain factors of these circuits are shown to be negative, representing the out-of-phase relationship of V_o and V_s, where V_s is measured at input I with respect to input II.

5.4-1   A Buffered Instrumentation Amplifier

The circuit shown in Fig. 5-7a is simply a differential-mode amplifier preceded by a voltage follower at each input. These voltage followers (buffers) have extremely large input impedances (resistances) and therefore provide effective isolation between the bridge and the differential-mode amplifier. Since each voltage follower's gain $A_v = 1$, the gain of the entire circuit is the

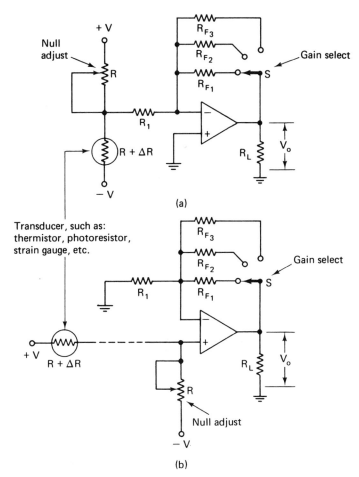

Figure 5-6   Half-bridge working into (a) an inverting circuit and (b) a noninverting circuit.

gain of the output differential stage. Thus, for the circuit shown in Fig. 5-7a,

$$\frac{V_o}{V_s} \cong \frac{R_F}{R_1}. \qquad (3\text{-}4)$$

5.4-2   A Two-Op Amp Instrumentation Amplifier

The circuit in Fig. 5-7b also has high input impedance due to a noninverting-mode amplifier at each input. This circuit's differential gain V_o/V_s is deter-

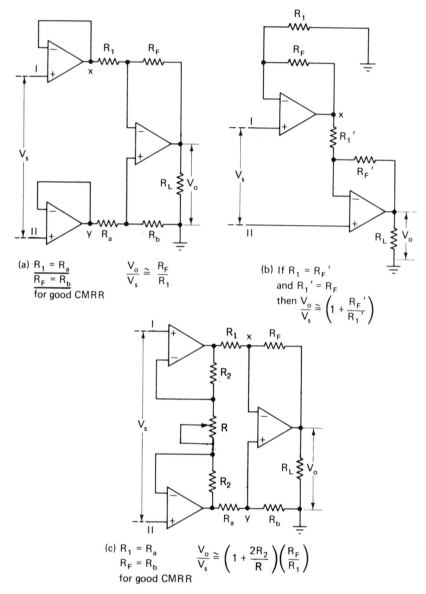

Figure 5-7   High input impedance differential-mode amplifiers.

mined by the components connected on the output (lower) Op Amp; that is

$$\frac{V_o}{V_s} \cong \left(1 + \frac{R'_F}{R'_1}\right). \qquad (5\text{-}9)$$

These voltage gains are negative (inverting) if V_s is referenced at input I with respect to input II. Once the gain and components R'_1 and R'_F are selected; the components in the upper stage must comply with the equations

$$R_F = R'_1$$

and

$$R_1 = R'_F.$$

5.4-3   Variable Gain Instrumentation Amplifier

A useful instrumentation amplifier is shown in Fig. 5-7c. It has high input resistance, a good CMRR, and variable gain that is adjustable with the potentiometer R. Generally, smaller values of R provide larger gains. The equation for gain is easy to derive. Note in Fig. 5-8 that because $V_{id} \cong 0$ on both buffers, the input signal V_s must appear across the potentiometer R. Also we note that the output signal V_{o_1} of these buffers is across both resistors R_2 and the potentiometer R. These carry the same current and can be viewed as series resistors. By Ohm's law, then,

$$\frac{V_o}{V_s} = \frac{(R_2 + R + R_2)I}{RI} = \frac{(2R_2 + R)I}{RI},$$

where I is the signal current that cancels to show that these buffers have a gain of

$$\frac{V_{o_1}}{V_s} = \frac{(2R_2 + R)}{R} = \frac{2R_2}{R} + 1.$$

The differential stage, following the buffers, has a gain of R_F/R_1. This,

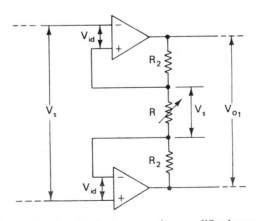

Figure 5-8   Buffer stage for instrumentation amplifier has variable gain.

combined with the gain of the buffers, yields a total gain of

$$\frac{V_o}{V_s} \cong \left(1 + \frac{2R_2}{R}\right)\left(\frac{R_F}{R_1}\right).$$

(5-10)

5.4-4   Simple One-Op Amp Instrumentation Amplifier

Another variable-gain differential-mode amplifier is shown in Fig. 5-9. Its gain V_o/V_s is varied by an adjustment of the potentiometer R_3, though its CMRR is affected thereby. Generally, the gain is increased or decreased by a decrease or increase of R_3, respectively. In this circuit, the right end of the feedback resistor R_F is connected to a voltage divider and the output signal voltage V_o is across this divider. When R_3 is maximum, half as much signal is fed back to the inverting input via R_F than would be the case if R_F were connected directly to the output as in the simpler differential amplifier in Fig. 5-3a. Thus the minimum gain can be estimated with the equation:

$$\frac{V_o}{V_s} \cong 2\left(\frac{R_F}{R_1}\right).$$

(5-11)

This shows that twice as much gain occurs with half as much negative feedback. If R_3 is adjusted to a minimum of 0 Ω, the right end of R_F is grounded and no negative feedback occurs. This pulls the stage gain up to

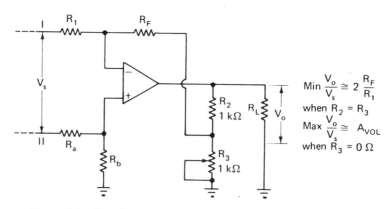

Figure 5-9   One Op Amp differential amplifier with variable gain.

the open-loop gain A_{VOL} of the Op Amp. Of course, both inputs can be preceded with voltage followers if high input impedance is required.

REVIEW QUESTIONS 5

5-1. What is a common-mode voltage?

5-2. What is the ideal value of CMRR?

5-3. If an Op Amp is to work in an environment that has significant induced voltages, does grounding *one* of the inputs reduce noise in the output? Why?

5-4. What do the initials CMRR mean?

5-5. Graphically, convert the following CMRR values to decibels: 10, 100, 1000, 10,000, 100,000, and 10^6.

5-6. Use your calculator to convert the following values of CMR (dB) to voltage ratios: 6 dB, 15 dB, 32 dB, 50 dB, 68 dB, 90 dB, and 110 dB.

5-7. What is the ideal value of common-mode voltage gain?

5-8. In what configuration is the Op Amp arranged to make induced noise voltages common mode?

5-9. What effect do the dc supply voltage values have on the maximum recommended common-mode voltages?

5-10. Assume that in the circuit of Fig. 5-3a, R_F is replaced with a potentiometer, while all other resistors remain fixed. What will happen to the circuit's differential gain A_v and its CMRR capability if the potentiometer is varied?

5-11. What is the purpose of using a voltage follower in each input of a differential amplifier?

PROBLEMS 5

Sections 5.1 and 5.2

5-1. In the circuit of Fig. 5-10, the output V_{cmo} measures 4 mV rms, 60 Hz, when the applied $V_{cm} = 400$ mV rms, 60 Hz. What is the Op Amp's common-mode rejection in decibels.

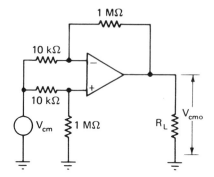

Figure 5-10   Circuit for Problems 5-1 and 5-2.

5-2. In the circuit of Fig. 5-10, the output V_{cmo} measures about 127 μV rms, 60 Hz, when the applied V_{cm} = 400 mV rms, 60 Hz. What is this Op Amp's common-mode rejection in decibels?

5-3. In the circuit of Fig. 5-11, if the signal voltage at input I varies from −30 mV up to 20 mV and if 10 mV rms, 60 Hz, is present at input I to ground, what are (a) the signal variations at output 0 and (b) the 60-Hz rms voltage V_n at output 0? The Op Amp's CMRR = 10,000.

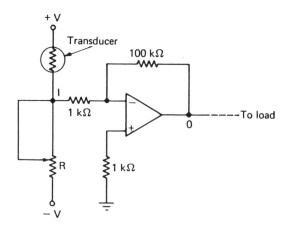

Figure 5-11   Circuit for Problems 5-3 and 5-4.

5-4. If the signal at input I to ground in the circuit of Fig. 5-11 varies from −25 mV to +35 mV while 15 mV rms, 60 Hz is also present at this input, what are (a) the signal variations and (b) the 60-Hz rms voltage V_n at output 0? The Op Amps's CMRR = 15,000.

5-5. Refer to the circuit in Fig. 5-12. If $R_F = R_b = 10$ kΩ, $R_1 = R_a = 1$ kΩ, CMR (dB) = 80 dB, signal across inputs I and II varies between -0.5 V and $+0.5$ V, and if both of these inputs have $+2$-V dc and 20-mV rms, 60 Hz, potentials with respect to ground, find (a) the approximate dc voltages to ground at inputs I and 2, (b) the output signal variations, and (c) the rms value of 60-Hz voltage at output 0.

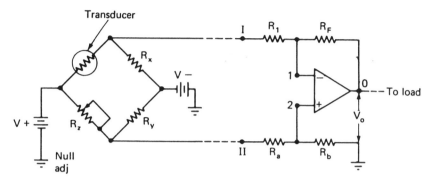

Figure 5-12   Circuit for Problems 5-5–5-9, 5-11, and 5-13.

5-6. If the circuit of Fig. 5-12 $R_F = R_b = 20$ kΩ, $R_1 = R_a = 500$ Ω, CMRR = 50,000, signal across inputs I and II varies from -80 mV to $+80$ mV, and if both of these inputs have $+4$-V dc and 600-mV rms, 60 Hz, potentials to ground, find (a) the approximate dc voltages to ground at inputs 1 and 2, (b) the output signal variations, and (c) the rms value of the 60-Hz voltage at output 0.

5-7. Suppose that in the circuit of Fig. 5-12, the supply voltages on the bridge and Op Amps are ± 15 V, $R_F = R_b = 10$ kΩ, $R_1 = R_a = 1$ kΩ and $R_y = 4$ kΩ. The Op Amp's output becomes 0 V (nulled) when R_z is adjusted to 1 kΩ. Find: (a) the dc common-mode input voltage, and (b) the dc to ground voltage at inputs 1 and 2 of the Op Amp.

5-8. Referring to the circuit described in the previous problem and to the Commercial Selection Guide of Appendix E, if the Op Amp is a LF13741, has its maximum common-mode input voltage been exceeded? Explain.

5-9. In the circuit of Fig. 5-12, $R_x = R_y = 1$ kΩ and a null is obtained when R_z is adjusted to 1 kΩ. The bridge and Op Amp supply voltages are ± 15 V, $R_1 = R_a = 110$ kΩ, and $R_F = R_b = 220$ kΩ. What is (a) the dc common-mode input voltage and (b) the range of output voltage V_o if the transducer's resistance varies over the range 900 Ω to 1.1 kΩ?

Section 5.3

5-10. In the circuit of Fig. 5-13, $R_y = 3.3$ kΩ, $R_z = 2.2$ kΩ, and a null is obtained when R_x is adjusted to 300 Ω. The bridge supply voltage $E = 5$ V and the Op Amp's supply voltages are ± 15 V. If $R_1 = R_a = 1.1$ kΩ and $R_F = R_b = 3.3$ kΩ, what are (a) the dc common-mode input voltage and (b) the range of output voltage V_o if the transducer's resistance varies over the range 190–210 Ω?

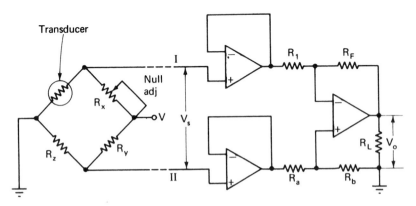

Figure 5-13 Circuit for Problems 5-10, 5-12, 5-13, and 5-14.

Review

5-11. Referring to the circuit described in Problem 5-9, over what range will the voltage at point I with respect to point II vary if the transducer's resistance varies from 0 Ω to infinity?

5-12. Referring to the circuit described in Problem 5-10, over what range will the voltage V_s vary if the transducer's resistance varies from 0 Ω to infinity?

5-13. In Problem 5-9, if V_o clips at ± 12 V, what maximum ratios of R_F/R_1 and R_b/R_a can we use and still avoid such clipping?

5-14. In Problem 5-10, if V_o clips at ± 12 V, what maximum ratios of R_F/R_1 and R_b/R_a can we use and still avoid such clipping?

Section 5.4

5-15. What is the gain V_o/V_s of the circuit in Fig. 5-14?

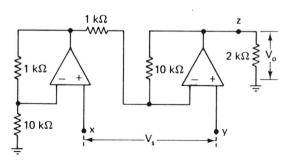

Figure 5-14   Circuit for Problems 5-15–5-18.

5-16. What is the gain V_o/V_s of the circuit in Fig. 5-14 if both 1-kΩ resistors are replaced with 2.2-kΩ values?

5-17. In the circuit of Fig. 5-14, if the voltage at point x is $+5.5$ V to ground and the voltage at y is $+5.1$ V to ground, what is the voltage at point z with respect to ground, assuming that the output was initially nulled? What is the average dc common-mode input voltage?

5-18. In the circuit of Fig. 5-14, if the voltages at points x and y are -0.1 V and -0.5 V, respectively, what is the voltage at point z with respect to ground, and what is the dc common-mode input voltage? Assume that the output was initially nulled.

5-19. Over what range can the gain V_o/V_s in the circuit of Fig. 5-15 be adjusted?

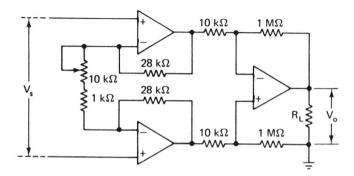

Figure 5-15   Circuit for Problems 5-19–5-22.

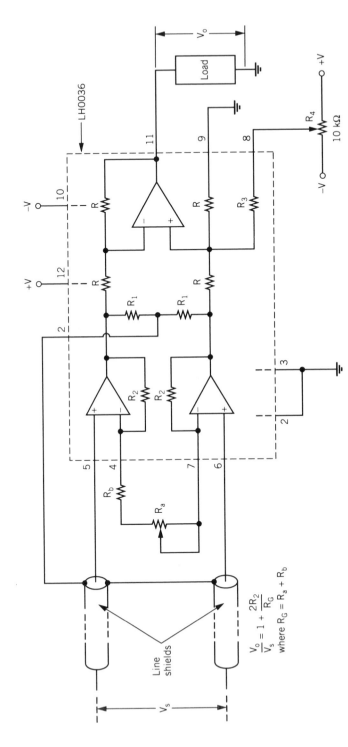

Figure 5-16 Typical application of an instrumentation amplifier LH0036, circuit for Problems 5-25–5-28.

$$\frac{V_o}{V_s} = 1 + \frac{2R_2}{R_G}$$

where $R_G = R_a + R_b$

5-20. If the 1-MΩ resistors and the 28-kΩ resistors are replaced with 10-kΩ values in the circuit of Fig. 5-15, over what range can its gain be adjusted?

5-21. What is the gain V_o/V_s of the circuit in Fig. 5-15 if its 1-kΩ resistor becomes open?

5-22. In the circuit of Fig. 5-15, if the 1-kΩ resistor is replaced with a short, to what theoretical maximum can this circuit's voltage gain be adjusted?

5-23. In the circuit of Fig. 5-9, $R_1 = R_a = 10$ kΩ, and $R_F = R_b = 22$ kΩ. Find its gain V_o/V_s when (a) R_3 is adjusted to 0 Ω, and when (b) R_3 is adjusted to 1 kΩ.

5-24. Referring to the circuit and components described in the previous problem, if inadvertently the fixed 1-kΩ resistor and 1-kΩ potentiometer are wired in each other's places, what is V_o/V_s when (a) the potentiometer resistance is 0 Ω and when (b) its resistance is 1 kΩ?

The remaining questions refer to the circuit of Fig. 5-16. The resistors R_1 in the LH0036 have no effect on the differential gain V_o/V_s of this circuit.

5-25. What are the purposes of the potentiometers R_a and R_4?

5-26. If $R_a = 0$ Ω, $R_b = 200$ Ω, and $R_2 = 25$ kΩ, what is the gain of this circuit?

5-27. If $R_a = 20$ kΩ, $R_b = 200$ Ω, and $R_2 = 25$ kΩ, what is the gain of this circuit?

5-28. If R_a is replaced with an open, what is the gain of this circuit?

6

THE OP AMP'S BEHAVIOR
AT HIGHER FREQUENCIES

Some of the characteristics of the practical Op Amp are sensitive to changes in operating frequency. In this chapter, we will consider problems caused by the decrease (roll-off) of the open-loop gain at higher frequencies. We will also see how the gain vs. frequency curve of an Op Amp can be tailored either by the circuit designer who uses the Op Amp or by its manufacturer. Other notable frequency-related problems such as reduced output-voltage swing capabilities, limited slew rates, and noise are also described here.

6.1 GAIN AND PHASE SHIFT VS. FREQUENCY

Ideally, an Op Amp should have an infinite bandwidth. This means that, if its open-loop gain is 90 dB with dc signals, its gain should remain 90 dB through audio and on to high radio frequencies. The practical Op Amp's gain, however, decreases (rolls off) at higher frequencies as shown in Fig. 6-1. This gain roll-off is caused by capacitances within the Op Amp circuitry. The reactances of these capacitances decrease at higher frequencies, causing shunt signal current paths, and thus reducing the amount of signal available at the output terminal. Along with decreased gain at higher frequencies, there is an increased phase shift of the output signal with respect to the input (see Fig. 6-2). Normally at low frequencies, there is a 180° phase difference in the signals at the inverting input and output terminals. But at higher frequencies, the output signal lags by more than 180°, and this is called *phase shift*. Thus the Op Amp with the characteristics in Fig. 6-2 has practically no phase shift up to about 30 kHz. Beyond 30 kHz, the output signal starts to lag, and at 300 kHz it lags an additional 40°. This negative or

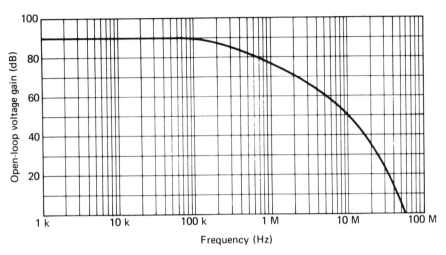

Figure 6-1   Typical open-loop gain vs. frequency curve of an Op Amp.

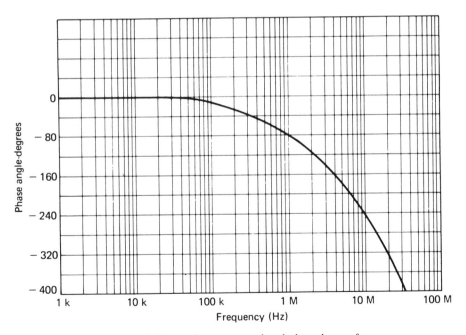

Figure 6-2   Typical open-loop output-signal phase lag vs. frequency.

lagging shift adds to the initially lagging 180°, causing an output signal lag of 220° compared with the signal applied to the inverting input. Similarly, at 1 MHz, the output lags the inverting input's signal by an additional 80°, or a 260° total. Problems caused by this kind of output-signal phase shift are discussed later in this chapter.

6.2 BODE DIAGRAMS

The gain and phase lag curves are often approximated with a series of straight lines.* Such straight-line approximations of gain and phase shift vs. frequency, where frequency is plotted on a logarithmic scale, are called *Bode diagrams* or *Bode plots*. A straight-line approximation of the curve in Fig. 6-1 is shown in Fig. 6-3. Note that, according to the approximated curve, this Op Amp's frequency response is flat from low frequencies (including dc) to 200 kHz. This means that its gain is constant from zero to 200 kHz, and therefore the bandwidth BW is about 200 kHz. Note that at higher frequencies, from 200 kHz to 2 MHz, the gain drops from 90 to 70 dB, which is at a -20-dB/decade† or -6-dB/octave rate. The negative sign refers to the negative (decreasing) slope of the curve. At frequencies from 2–20 MHz, the roll-off is -40 dB/decade or -12 dB/octave. About 20 MHz, the roll-off is -60 dB/decade or -18 dB octave.

Because Op Amps are seldom used in open loop for amplification of signals, we must consider the effects of feedback on the Op Amp's frequency response. Figure 6-4 shows that if the Op Amp is wired to have a closed-loop gain $A_v = 10,000$ or 80 dB, the bandwidth is 600 kHz. We determine this by projecting to the right of 80 dB to the point where this projection intersects the open-loop curve. Directly below this point of intersection, we read 600 kHz on the horizontal scale. Note that the open-loop curve is descending at -20 dB/decade at the point where the 80-dB projection intersects it. We can, therefore, say that the curve and projection intersect at a 20-dB/decade rate of closure.

If the Op Amp is wired for a closed-loop gain $A_v = 1000$ or 60 dB, we project to the right of 60 dB toward the open-loop curve. In this case, the projection and curve intersect at a 40-dB/decade rate of closure and at a point above 3.5 MHz, approximately. The bandwidth *appears* to be about 3.5 MHz. However, because the open-loop curve is descending at -40 dB/decade, the circuit will very likely be *unstable* and should not be used without

---

*The straight lines are linear sums of the asymptotes on the individual stage gain vs. frequency curves.

†A -20-dB/decade gain roll-off means that the open-loop gain decreases 20 dB with a frequency increase by the factor 10, and is equivalent to a -6-dB/octave roll-off. A frequency change by an octave is a change by the factor 2.

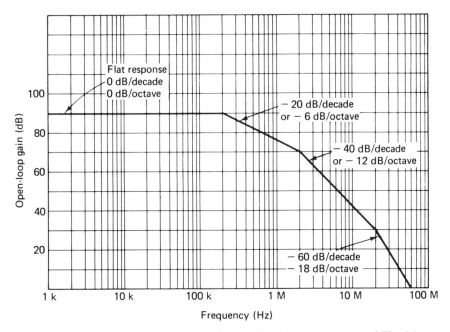

Figure 6-3 Approximation of open-loop gain vs. frequency curve of Fig. 6-1.

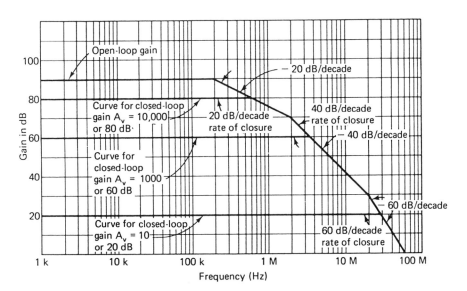

Figure 6-4

modifications. Oscillation is a symptom of an unstable amplifier and is discussed further in later chapters. With Op Amps specifically, the main causes of instability are:

1. In the -40-dB/decade roll-off region, the Op Amp's output signal phase lag is so large that it becomes, or approaches, an in-phase condition with the signal at the inverting input.
2. With lower closed-loop gains, the feedback resistor R_F is relatively small compared to resistance R_1 (see Figs. 3-5 and 3-10). This increases the amount of signal that is fed back to the inverting input.

The effect of the output signal becoming more in phase with the signal normally at the inverting input, combined with the greater amount of this output being fed back, makes the Op Amp oscillate at higher frequencies when wired to have relatively low closed-loop gain. In other words, at higher frequencies and lower closed-loop gains the feedback becomes significant and regenerative and thus meets the requirements of an oscillator.

Generally, the rate of closure between the closed-loop gain projection and the open-loop curve should not significantly exceed 20 dB/decade or 6 dB/octave for stable operation. Therefore the Op Amp with the characteristics in Fig. 6-4 is likely to be unstable if wired for gains below about 70 dB. Thus, as mentioned before, if this Op Amp is wired for 60-dB gain, the rate of closure is 40 dB/decade, and the circuit will probably be unstable. Also as shown in Fig. 6-4, a gain of 20 dB causes a 60-dB/decade rate of closure, which also means unstable operation is likely. When unstable, the circuit can have unpredictable output signals even if no input signal is applied.

6.3   EXTERNAL FREQUENCY COMPENSATION

Some types of Op Amps are made to be used with externally connected compensating components—capacitors and sometimes resistors—if they are to be wired for relatively low closed-loop gains. These are called *uncompensated* or *tailored-response* Op Amps because the circuit designer must provide the compensation if it is required. The compensating components alter the open-loop gain characteristics so that the roll-off is about 20 dB/decade over a wide range of frequencies. For example, Fig. 6-5 shows some gain vs. frequency curves that are obtained with various compensating components on a tailored-response Op Amp. In this case, if $C_1 = 500$ pF, $C_2 = 2000$ pF, and $C_3 = 1000$ pF, then the gain vs. frequency curve 1 applies according to the table, and the roll-off starts at about 1 kHz with a steady -20-dB/decade rate. Thus if feedback components are now added to obtain an 80-dB closed-loop gain, the circuit's frequency response is flat from dc to about 3 kHz. This can be seen if we project to the right of 80 dB to curve 1. The

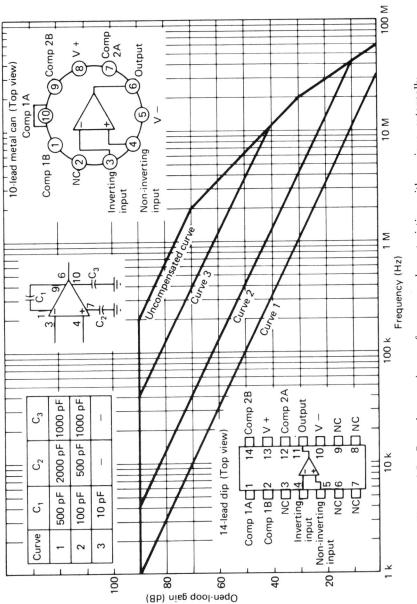

Figure 6-5 Open-loop gain *vs.* frequency characteristics with various externally connected (outboard) compensating components.

Curve	C_1	C_2	C_3
1	500 pF	2000 pF	1000 pF
2	100 pF	500 pF	1000 pF
3	10 pF	—	—

intersection of this projection and curve 1 is at 3 kHz. Better bandwidth is obtained with curve 1 if a lower closed-loop gain is used; that is, if $A_v = 100$ or 40 dB, the bandwidth increases to about 300 kHz. In either case, the rate of closure is 20 dB/decade, and the circuit is stable. Generally, if low gain is required, compensating components for curve 1 should be used. If high gain and relatively wide bandwidth are required, the compensating component for curve 3 should be used. Note in the table that if curve 3 is used, only one compensating component, $C_1 = 10$ pF, is required in this case. No compensating components need be used if this Op Amp is wired for more than 70 dB because the rate of closure never exceeds 20 dB/decade with such high gains. Note also on curve 3 that a closed-loop gain of less than 30 dB causes this gain vs. frequency response curve (projection) to meet the open-loop gain curve at a point where the rate of closure is greater than 20 dB/decade; therefore, a circuit with this gain is likely to be unstable.

Example 6-1

If the data in Fig. 6-5 apply to the Op Amp in Fig. 6-6, what are this circuit's gain V_o/V_s and bandwidth with each of the following switch positions: I, II, and III? Is this circuit stable in each switch position?

Answer. This is a noninverting circuit, and therefore its gain is

$$A_v = V_o/V_s \cong \frac{R_F}{R_1} + 1. \qquad (3\text{-}6)$$

The compensating capacitors give us the gain vs. frequency curve 2.

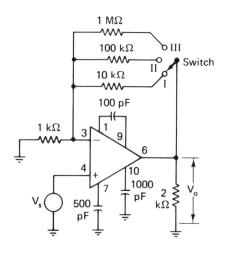

Figure 6-6 For Example 6-1.

With the switch in position I,

$$A_v \cong \frac{10 \text{ k}\Omega}{1 \text{ k}\Omega} + 1 = 11.$$

This is equivalent to about 20.8 dB or about 20 dB. Projecting to the right from 20 dB in Fig. 6-5, we intersect curve 2 at 13 MHz, which means that the bandwidth in this case is about 13 MHz.

With the switch in position II,

$$A_v \cong \frac{100 \text{ k}\Omega}{1 \text{ k}\Omega} + 1 = 101,$$

which is equivalent to about 40 dB. Projecting to the right of 40 dB we intersect curve 2 at about 1.3 MHz, and therefore the bandwidth now is about 1.3 MHz.

With the switch in position III,

$$A_v \cong \frac{1 \text{ M}\Omega}{1 \text{ k}\Omega} + 1 = 1001,$$

which is equivalent to about 60 dB. A projection to the right of 60 dB intersects curve 2 above 130 kHz, which means that the bandwidth now is 130 kHz.

Since all three projections intersect curve 2 at a -20-dB/decade rate of closure, the circuit is stable regardless of the switch position. Note that with these compensating components, if this Op Amp is wired as a voltage follower; that is, with a gain of less than 10 dB, it would be unstable.

6.4 COMPENSATED OP AMPS

Sometimes the relatively broad bandwidth of the uncompensated Op Amps is not needed. For example, in the instrumentation circuit shown in Fig. 5-5 of the previous chapter, the Op Amp is required to amplify relatively slowly changing signals, and therefore it does not require good high-frequency response. In this and similar applications, internally compensated Op Amps can be used. They are usually called *compensated* Op Amps. Also, they are stable regardless of the closed-loop gain and without externally connected compensating components.

The type 741 Op Amp is compensated and has an open-loop gain vs. frequency response as shown in Fig. 6-7. The IC of the 741 contains a 30-pF capacitance that internally shunts off signal current and thus reduces the available output signal at higher frequencies. This internal capacitance, which is an internal compensating component, causes the open-loop gain to

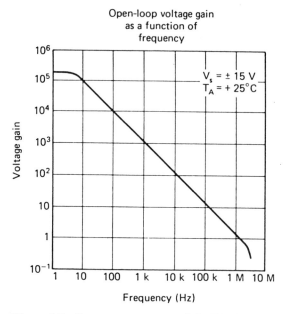

Figure 6-7   Frequency response of the 741 Op Amp.

roll off at a steady 20-dB/decade rate. Therefore, regardless of the closed-loop gain we use, the gain projection will intersect the open-loop gain curve at a 20-dB/decade rate of closure, and this assures us of a stable circuit.

The 741, along with other types of compensated Op Amps, has a 1-MHz *gain-bandwidth product*. This means that the product of the coordinates, gain and frequency, of any point on the open-loop gain vs. frequency curve is about 1 MHz.

Obviously, if the 741 Op Amp is wired for a closed-loop gain of 10^4 or 80 dB, its bandwidth is 100 Hz, as can be seen by projecting to the right from 10^4 to the curve in Fig. 6-7. If the closed-loop gain is lowered, say to 10^2 or to 1, the bandwidth increases to 10 kHz or 1 MHz, respectively. The fact that the 741 has a 1-MHz bandwidth with unity gain explains why the 741 is listed with a *unity gain-bandwidth* of 1 MHz on typical specification sheets.

Example 6-2

If the Op Amp in the circuit of Fig. 6-8 is a type 741 and the circuit is required to amplify signals up to about 10 kHz, what maximum value of POT resistance can we use and still keep the circuit's response flat to 10 kHz?

Answer. The frequency response curve in Fig. 6-7 shows that the bandwidth is greater than 10 kHz if the closed-loop gain is kept under 100. Therefore,

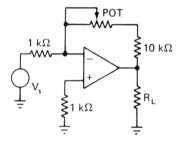

Figure 6-8 For Example 6-2.

we want to keep the ratio R_F/R_1 under 100. Thus since

$$\frac{R_F}{R_1} < 100, \quad \text{then} \quad R_F < 100R_1 = 100(1 \text{ k}\Omega) = 100 \text{ k}\Omega.$$

If we must keep the total feedback resistance R_F under 100 kΩ, then the POT's maximum resistance is 90 kΩ. Note the 10-kΩ fixed resistor in series with it.

6.5 SLEW RATE

An Op Amp's slew rate is related to its frequency response. Generally, we can expect Op Amps with wider bandwidths to have higher (better) slew rates. The slew rate, as mentioned in Chapter 2, is the rate of output-voltage change caused by a step input voltage and is usually measured in volts per microsecond. An ideal slew rate is infinite, which means that the Op Amp's output voltage should change instantly in response to an input step voltage. Practical IC Op Amps have specified slew rates from 0.1 V/μs to about 100 V/μs which are measured in special circumstances. Some hybrid* Op Amps have slew rates on the order of 1000 V/μs. Unless otherwise specified, the slew rate listed in a data sheet was probably measured with unity gain and open load.

A less than ideal slew rate causes distortion especially noticeable with nonsinusoidal waveforms at higher frequencies. For example, ignoring over-shoot, Fig. 6-9 shows typical output waveforms of a voltage follower with square waves of various frequencies applied. In this case, the slew rate is 1 V/μs. When the input frequency is 100 Hz with the waveform in Fig. 6-9b, the output has the waveform shown in part c of this figure. Similarly, when the input is 10 kHz or 1 MHz with the waveform shown in b, the output has the waveform in Fig. 6-9d or e, respectively. Obviously, due to the limited

---

*Hybrids, as opposed to monolithics (made on one chip), might contain one or more ICs or ICs and discrete components.

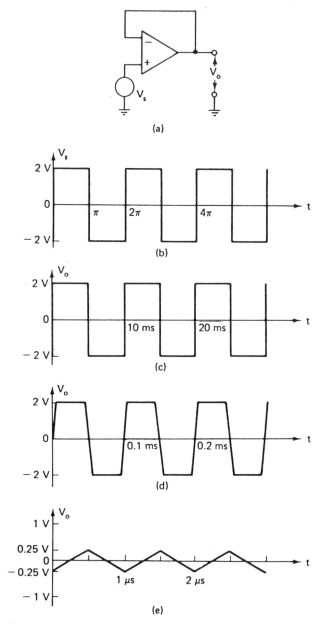

Figure 6-9 Voltage follower with an Op Amp having a slew rate of 1 V/μs and square-wave voltage applied. (c) Output when the frequency of V_s is 100 Hz. (d) Output when the frequency of V_s is 10 kHz. (e) Output when the frequency of V_s is 1 MHz.

slew rate, the square-wave input is distorted into a sawtooth at higher frequencies.

6.6 OUTPUT SWING CAPABILITY VS. FREQUENCY

An important consideration is an Op Amp's peak-to-peak output signal capability at higher frequencies. Generally, this capability decreases with higher frequencies, as shown in Fig. 6-10. In Fig. 6-10a, an uncompensated

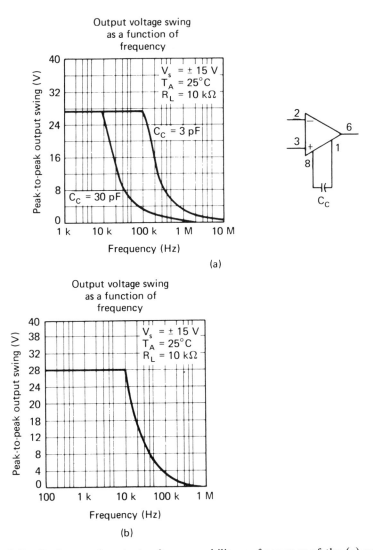

(a)

(b)

Figure 6-10 Peak-to-peak output voltage capability vs. frequency of the (a) uncompensated type 777 Op Amp; (b) compensated type 741 Op Amp.

Op Amp's characteristics are shown for two different compensating capacitors C_c. Note that if $C_c = 30$ pF is used, the peak-to-peak output capability decreases sharply at operating frequencies above 10 kHz. On the other hand, when $C_c = 3$ pF is wired externally (outboard), the bandwidth increases, as it did with smaller compensating capacitors shown in Fig. 6-5. Also, the peak-to-peak output capability does not drop until frequencies on the order of 100 kHz or more are applied.

The output vs. frequency curve in Fig. 6-10b is for the type 741 Op Amp, which is internally compensated. Apparently then, internal compensation not only limits bandwidth, but it also limits the peak-to-peak output capability. We must keep in mind, though, that internal compensation has outstanding advantages for instrumentation and low-frequency work. For example, it simplifies our work by not forcing our attention to externally wired compensation and by assuring us of stable operation.

Example 6-3

If the Op Amp in the circuit of Fig. 6-8 is a type 741 with dc source voltages of ± 15 V, $R_L = 10$ kΩ, and the POT adjusted to zero, with which of the following audio-frequency input signals will this circuit clip the output signals? (Assume that the Op Amp was initially nulled and that the input signals have no dc component.)

(a) 1 Hz to 10 kHz with peaks up to ± 1 V;

(b) 1 Hz to 100 kHz with peaks up to ± 1 V;

(c) 1 Hz to 100 kHz with peaks up to ± 10 mV.

Answer. With the POT resistance 0 Ω, the total feedback resistance is 10 kΩ. Since this is an inverting amplifier, its gain $A_v = -R_F/R_1 = -10$ k$\Omega/1$ k$\Omega = -10$. With this gain, the 741's bandwidth is about 100 kHz. See Fig. 6-7. All three signals appear to be within the bandwidth capability of this circuit. Therefore, we can focus our attention to the peak-to-peak output capability vs. frequency.

(a) With the 1-Hz to 10-kHz input signal peaking to ± 1 V, the output will peak to ± 10 V. Since the 741 can handle from 28 V peak to peak (14-V peak) up to 10 kHz, no clipping occurs.

(b) With the 1-Hz to 100-kHz input signal peaking up to ± 1 V, the output *attempts* to peak at ± 10 V over this frequency range. However, according to the curve in Fig. 6-10b, this output is clipped with signal frequencies above 15 kHz approximately. Projecting to the right of 20 V peak to peak (10-V peak), we intersect the curve at roughly

15 kHz. Thus input signals with peaks of 1 V are clipped more or less, depending on how far above 15 kHz the signal is.

(c) With the 1-Hz to 100-kHz input signal peaking up to ± 10 mV, the output peaks at ± 100 mV. No clipping occurs because the 741's peak-to-peak output capability is about 2-V or 1-V peak up to 100 kHz. This capability is well over the ± 100-mV peaks we intend to get.

6.7 HARMONICS IN NONSINUSOIDAL WAVEFORMS

The signals processed by Op Amps often are nonsinusoidal waveforms. If such waveforms are repetitious, they can be shown to consist of some *fundamental* sine-wave frequency and one or more of its *harmonics*. Harmonics are also sine waves with frequencies that are *integer* multiples of the fundamental. For example, if a nonsinusoidal waveform repeats every 1 ms, its fundamental frequency $f = 1/T = 1/1$ ms $= 1$ kHz. Its second harmonic is 2 kHz, its third harmonic is 3 kHz, etc. Examples of nonsinusoidal waveforms and their harmonic content are shown in Fig. 6-11.

The waveforms b through d of Fig. 6-11 are of particular interest to us. Note in these cases that the *resulting* nonsinusoidal waveforms V_R are somewhat square and that they consist of *odd*-numbered harmonics only. Especially note that the greater the number of odd harmonics, the squarer the resulting voltage V_R is. Generally, squarer waveforms contain *more* odd-numbered harmonics.

Referring back to Fig. 6-9, we observed that the applied voltage waveform b is squarer than the output waveform d. This output signal d, therefore, contains fewer higher-frequency harmonics that does the input signal b. This is caused by the fact that an Op Amp has a limited bandwidth and cannot pass high frequencies. Thus, the higher-frequency harmonics are removed or attenuated resulting in the less square output.

Amplifiers can either remove or add harmonics from or to the signals being amplified. Fig. 6-12 shows input signals and resulting outputs of three Op Amp circuits. In circuit a, the output V_o is a phase-inverted version of the input V_s but otherwise is an undistorted signal. No significant harmonics have been added or removed. The circuit b has a square-wave input V_s but a less square output V_o. This output signal takes more time to increase (rise) and to decrease (fall), indicating that the signal has been stripped of its higher-frequency harmonics. This amplifier, therefore, has effectively removed higher-frequency components from the signal being amplified. The circuit c clips (distorts) the signal. Since the output is squarer than the input, this amplifier adds harmonics. The frequencies of the created harmonics are mainly odd-integer multiples of the input signal frequency.

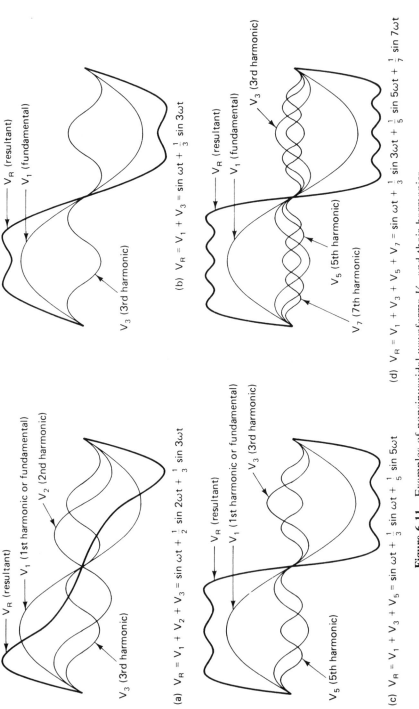

(a) $V_R = V_1 + V_2 + V_3 = \sin \omega t + \frac{1}{2} \sin 2\omega t + \frac{1}{3} \sin 3\omega t$

(b) $V_R = V_1 + V_3 = \sin \omega t + \frac{1}{3} \sin 3\omega t$

(c) $V_R = V_1 + V_3 + V_5 = \sin \omega t + \frac{1}{3} \sin \omega t + \frac{1}{5} \sin 5\omega t$

(d) $V_R = V_1 + V_3 + V_5 + V_7 = \sin \omega t + \frac{1}{3} \sin 3\omega t + \frac{1}{5} \sin 5\omega t + \frac{1}{7} \sin 7\omega t$

Figure 6-11 Examples of nonsinusoidal waveforms V_R and their harmonics.

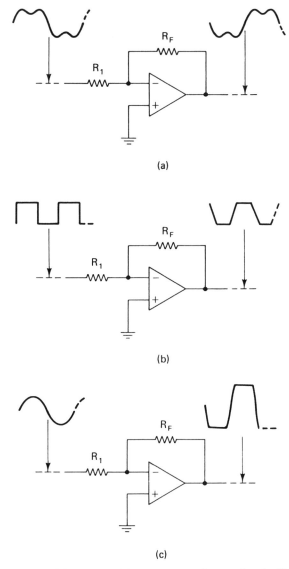

(a)

(b)

(c)

Figure 6-12   Amplifier (a) neither adds nor removes harmonics significantly; amplifier (b) removes higher-frequency harmonics; and amplifier (c) adds harmonics.

6.7-1   Rise and Fall Times of Nonsinusoidal Waveforms

The terms *rise time* T_R and *fall time* T_F are commonly used to describe the characteristics of nonsinusoidal waveforms. As shown in Fig. 6-13, the rise time T_R is defined as the time it takes a voltage to rise from 10% to 90% of its peak-to-peak amplitude V_{max}. Similarly, the fall time T_F is the time it takes the waveform to fall from 90% to 10% of V_{max}. These definitions

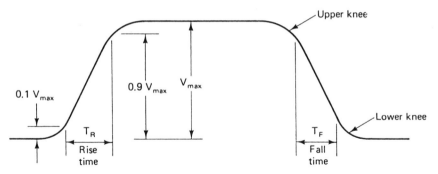

Figure 6-13 T_R is the time it takes the waveform to rise from 10% to 90% of its peak-to-peak amplitude V_{max}. T_F is the time it takes the waveform to fall from 90% to 10% of its peak-to-peak value V_{max}.

exclude the lower and upper *knees* of the waveform and, therefore, the T_R or T_F measurements are made in the most linear portions of the rise or fall.

6.7-2   Bandwidth vs. Rise and Fall Times

In previous sections we learned graphically or by brief calculations with an Op Amp's gain-bandwidth product how to determine bandwidths of Op Amp circuits. These are reliable methods but only with relatively small amplitude output signals. With larger output signals, an Op Amp's slew rate becomes a dominating factor causing the bandwidth to become narrower than the graphical (small-signal) analysis indicates.

We can determine an Op Amp circuit's large-signal bandwidth BW two ways. One is by driving its input with a square wave and then observing the resulting output signal's rise or fall time. The *measured* T_F or T_F can then be used in the equation

$$\boxed{\text{BW} \cong \frac{0.35}{T_R} = \frac{0.35}{T_F}.}\qquad (6\text{-}1)^*$$

The second method is to calculate the rise or fall time using the Op Amp's specified slew rate. The *calculated* T_R or T_F can then be used with the foregoing equation to determine the bandwidth BW.

*See Appendix I for derivation.

Example 6-4

A square wave, having an extremely fast rise and fall time, is applied to the input of an Op Amp circuit. Without saturating the Op Amp, the resulting output is a trapezoidal waveform, like that of Fig. 6-9d. By adjustment of the sweep time on a triggered sweep oscilloscope, we observe the leading edge of the trapezoid as shown in Fig. 6-14. If the SWEEP TIME control is in the CAL 10-μs/DIV position, what is this circuit's bandwidth?

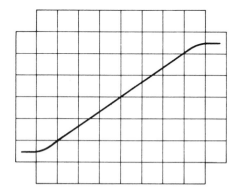

Figure 6-14 Rise time T_R measured on an oscilloscope.

Answer. The rise time T_R takes about six horizontal divisions. Since each division is 10 μs.

$$T_R = 6(10 \ \mu s) = 60 \ \mu s.$$

Using this measured T_R in Eq. (6-1) we find that the bandwidth

$$BW \cong \frac{0.35}{60 \ \mu s} \cong 5.8 \ \text{kHz}.$$

Example 6-5

A 741 is to output unclipped signals as large as 20 V peak to peak. What bandwidth can we expect with this output amplitude? The 741 has a 0.5 v/μs slew rate.

Answer. Since the 741 has a slew rate of about 0.5 V/μs, the total time T_{tot} it takes the output voltage to rise by 20 V is

$$T_{\text{tot}} = \frac{20 \ \text{V}}{0.5 \ \text{V}/\mu s} = 40 \ \mu s.$$

The rise time T_R is about 80% of this because the first and last 10% of T_{tot} is not included in the definition of T_R. In this case, then,

$$T_R \cong 0.8(40 \ \mu s) = 32 \ \mu s,$$

and, therefore, the bandwidth

$$BW \cong \frac{0.35}{32 \ \mu s} \cong 10.9 \ kHz. \tag{6-1}$$

6.8   NOISE

In Chapter 5 we referred to induced 60-Hz hum as noise. Indeed, any unwanted signal mixing with desired signals is noise. Induced noise is not limited to the 60 Hz from nearby electrical power equipment. It often originates in other man-made systems such as switch arcing, motor brush sparking, and ignition systems. Natural phenomena such as lightning also cause induced noise. Induced noise voltages, as shown in Chapter 5, can be made common mode and thus can be significantly reduced relative to the desired signals at the load of an Op Amp.

The term *noise* is also commonly used to describe ac random voltages and currents generated within conductors and semiconductors. Such noise, associated with Op Amps and with amplifiers in general, limit their signal sensitivity. If very weak signals are to be amplified, very high closed-loop gain must be used to bring the signals up to useful levels. With a very high gain, however, the noise is amplified along with the signals to the point where nearly as much noise as signal appears at the output. If fed to a speaker, random noise causes a hissing, frying sound.

There are three main types of noise phenomena associated with Op Amps and with solid-state amplifiers in general. These are: *thermal* or *Johnson* noise, *shot* or *Schottky* noise, and *flicker* or $1/f$ noise.

Thermal noise is caused by the random motion of charge carriers within a conductor which generate noise voltages V_n within it. The rms value of this thermally generated noise voltage V_n can be predicted with the equation

$$\boxed{V_n = \sqrt{4KTR(BW)} \,,} \tag{6-2}$$

where K is Boltzmann's constant; 1.38×10^{-23} joules/$^\circ$K,

T is the temperature in degrees Kelvin—the Celsius temperature plus 273°,

R is the resistance of the conductor in question, and

BW is the bandwidth in hertz.

Examining this equation, we see that thermal noise increases with higher temperatures, larger resistances, and wider bandwidths.

Although a constant average dc current may be maintained in a semiconductor, it will have random variations. These variations have an rms value referred to as noise current I_n. Noise generated in this manner is called *shot noise*. Its rms value can be predicted with the equation

$$I_n = \sqrt{2qI_{dc}(\text{BW})} \, ,$$
(6-3)

where q is the charge of an electron: 1.6×10^{-19} coulombs,

  I_{dc} is the average dc current in the semiconductor, and

  BW is the bandwidth.

Here again, wider bandwidths generate more noise, which is to say, we can expect less noise in narrow-bandwidth amplifiers. Of course, the noise current I_n, flowing through a resistance R, will generate noise voltage RI_n.

In addition to shot noise, semiconductors have low-frequency noise called *flicker* or $1/f$ noise. The term $1/f$ describes the inverse nature of this noise with respect to frequency; that is, the amount of flicker noise is greater at lower frequencies f.

6.9   EQUIVALENT INPUT NOISE MODEL

An Op Amp contains many active and passive components that generate and add noise to its output. These noise sources can be represented by voltage and current noise generators at the input of the amplifier's equivalent circuit, as shown in Fig. 6-15. This equivalent is sometimes called an *amplifier noise model*. The net effect of these input noise generators is an *equivalent*

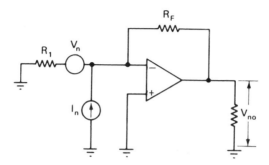

Figure 6-15 An amplifier's inherent noise voltage and current can be shown as voltage and current generators at its inputs.

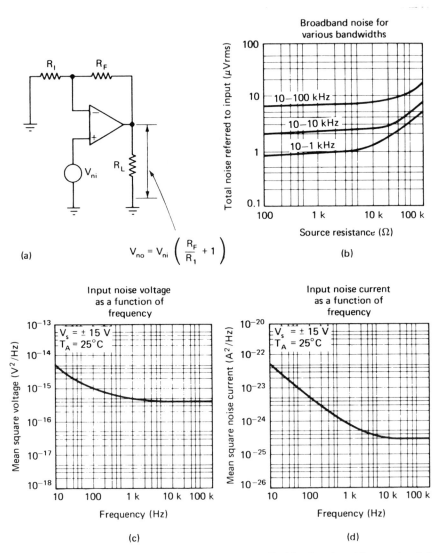

(a)

$$V_{no} = V_{ni} \left(\frac{R_F}{R_1} + 1 \right)$$

(b)

(c)

(d)

Figure 6-16 The total effective input noise is amplified by the closed-loop gain of the amplifier seeing it as a noninverting type. (b) Typical noise levels higher with wider bandwidths and with larger signal source resistances. (c) The rms noise voltages squared typically developed at various frequencies. (d) The rms noise currents squared typically developed at various frequencies.

input-noise voltage V_{ni}. It is this equivalent input noise V_{ni} that is amplified by the stage gain, and with it we can predict the output noise V_{no} that a given Op Amp stage might have.

If we neglect the thermal noise of the amplifier's signal source, the equivalent input noise V_{ni} is related to the amplifier's specified input noise voltage V_n and noise current I_n by the following equation:

$$V_{ni}^2 \cong V_n^2 + (R_E I_n)^2, \qquad (6\text{-}4)$$

where V_{ni}^2 is the mean-square equivalent input noise voltage.

V_n^2 is the specified mean-square input noise voltage.

I_n^2 is the specified mean-square input noise current, and R_E is the parallel equivalent of R_1 and R_F.

It can be shown that equivalent input noise voltage V_{ni} sees the Op Amp circuit as a noninverting type as shown in Fig. 6-16. Broadband input noise vs. source resistance curves are sometimes shown as in Fig. 6-16b, and the general-purpose Op Amp's specified noise voltage V_n and input noise current I_n vs. frequency characteristics are shown in Fig. 6-16c and d. Larger source resistances appreciably add to thermally generated noise as shown in part b of the figure. The $1/f$ phenomenon adds low-frequency noise as shown in Fig. 6-16c and d.

The output noise voltage V_{no} is a function of the effective input noise voltage V_{ni} and is approximated with the equation

$$V_{no}^2 \cong (A_v V_{ni})^2 \qquad (6\text{-}5a)$$

or

$$V_{no} \cong A_v \sqrt{V_n^2 + (R_E I_n)^2}, \qquad (6\text{-}5b)$$

where A_v is the closed-loop gain.

REVIEW QUESTIONS 6

6-1. Referring to the open-loop gain of an Op Amp, what does the term *roll-off* mean?

6-2. What is a Bode diagram?

6-3. What problem can occur with an amplifier having a steep, greater than a -20-dB/decade, gain vs. frequency roll-off?

6-4. How can the steepness of the gain vs. f roll-off deliberately be limited?

6-5. Why are low closed-loop gains avoided with uncompensated Op Amps?

6-6. What advantage does the constant -20-dB/decade roll-off of the internally compensated Op Amp have?

6-7. What disadvantage does the internally compensated Op Amp have compared to the externally compensated types?

6-8. If an Op Amp is specified as having a *tailored response*, what does this mean?

6-9. If an Op Amp is specified as being *compensated*, what does this mean?

6-10. If the closed-loop gain of an Op Amp is increased by increasing its feedback resistance R_F, what effect will this have on its bandwidth?

6-11. If an Op Amp's gain-bandwidth product is 2 MHz, what is its bandwidth when connected to work as a voltage follower?

6-12. What does the term slew rate mean?

6-13. What effect does the operating frequency have on the maximum unclipped output signal capability of an Op Amp?

6-14. When compensating components are added externally on an Op Amp, what effect do they have on the bandwidth?

6-15. Name the three types of noise associated with conductors and semiconductors.

6-16. Generally, at what frequencies is flicker noise a greater problem?

6-17. What effect do large externally wired resistances have on the noise level at the output of an Op Amp?

PROBLEMS 6

Refer to Fig. 6-17a. The Op Amp will work on one of the various open-loop vs. frequency curves shown in Fig. 6-17c, depending on the values of the externally wired compensating components actually used.

Sections 6.1–6.3

6-1. For the Op Amp and characteristics of Fig. 6-17, what is the gain-bandwidth product if $C_1 = 5000$ pF, $R_1 = 1.5$ kΩ, and $C_2 = 200$ pF?

6-2. For the Op Amp and characteristics of Fig. 6-17, what is the gain-bandwidth product in the frequency range from 1 kHz to 1 MHz if $C_1 = 500$ pF, $R_1 = 1.5$ kΩ, and $C_2 = 20$ pF?

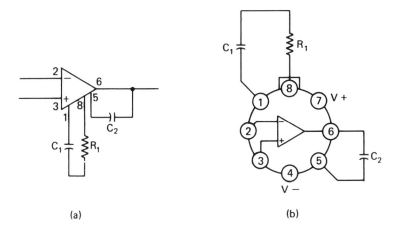

(a)   (b)

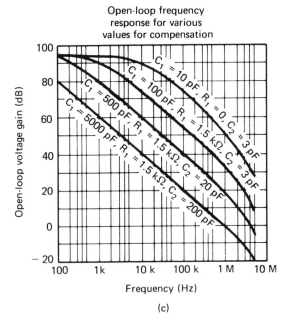

(c)

Figure 6-17   Op Amp with tailored response, and typical response curves; Problems 6-1–6-4.

6-3. What values of compensating components should we use with an Op Amp that has characteristics of Fig. 6-17 if we need a gain-bandwidth product of 10 MHz in the range of 40–60-dB closed-loop gain? *Hint:* Convert gain in decibels to gain as a voltage ratio.

6-4. If an Op Amp has a 10-MHz gain bandwidth product, what is its bandwidth when used with a closed-loop gain of (a) 40 dB, and when it is (b) 60 dB?

6-5. If the Op Amp in Fig. 6-18 has the characteristics given in Fig. 6-17, what is the circuit's bandwidth when the 100-kΩ POT is adjusted to 0 Ω?

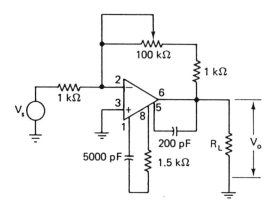

Figure 6-18   Problems 6-5, 6-6, and 6-11.

6-6. If the Op Amp in Fig. 6-18 has the characteristics given in Fig. 6-17, what is the circuit's bandwidth when the 100-kΩ POT is adjusted to maximum?

6-7. What is the bandwidth of the circuit in Fig. 6-19 if its 1-MΩ POT is adjusted to 900 kΩ and if its Op Amp has the characteristics given in Fig. 6-17?

6-8. What is the bandwidth of the circuit in Fig. 6-19 if its 1-MΩ POT is adjusted to 0 Ω and if the Op Amp has characteristics given in Fig. 6-17?

Section 6.4

6-9. What are the approximate gain and bandwidth of the circuit in Fig. 6-20 if the 1-kΩ POT is adjusted to 0 Ω and the Op Amp has a gain-bandwidth product of 1 MHz?

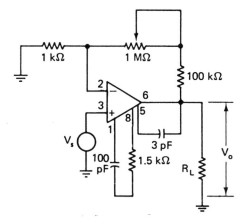

Figure 6-19 Problems 6-7 and 6-8.

6-10. If the 1-kΩ POT in the circuit of Fig. 6-20 is adjusted for maximum resistance and if the Op Amp has a gain-bandwidth product of 1 MHz, what are this circuit's gain and bandwidth?

6-11. If the Op Amp in the circuit of Fig. 6-18 has the characteristics in Fig. 6-17, what is the rate of closure between the closed-loop and the open-loop gains vs. frequency curves regardless of the adjustment of the 100-kΩ POT?

6-12. In Fig. 6-20, if the Op Amp is a 741, what is the rate of closure between the closed-loop and the open-loop gain vs. frequency curves with most adjustments of the 1-kΩ POT?

Section 6.6

6-13. The Op Amp circuit in Fig. 6-21a has the characteristics shown in Fig. 6-21b. Its input signal V_s has variable frequency and often peaks up to 80 mV. If $R_F = 100$ kΩ, beyond what applied frequency is the output signal V_o likely to be clipped? Assume that the circuit was initially nulled.

6-14. Referring to the Op Amp circuit and characteristics in Fig. 6-21, what maximum peak-to-peak voltage value of V_s can we apply and not clip the output signal V_o if $R_F = 200$ kΩ and the amplifier must have a flat response up to 100 kHz?

6-15. Referring to the Op Amp and its characteristics in Fig. 6-22, how much common-mode voltage can we expect across the load R_L if the input common-mode voltage $V_{cm} = 2$ mV rms at 60 Hz?

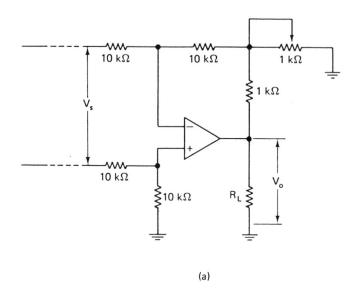

(a)

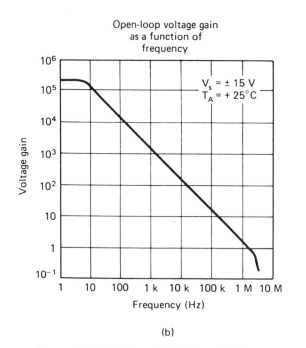

(b)

Figure 6-20   Problems 6-9, 6-10, and 6-12.

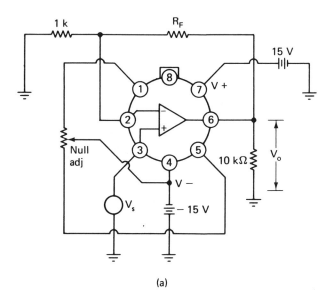

(a)

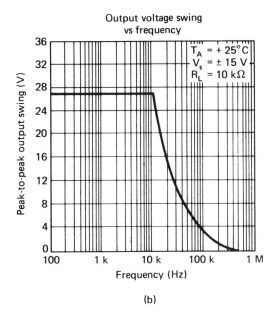

(b)

Figure 6-21   Problems 6-13 and 6-14.

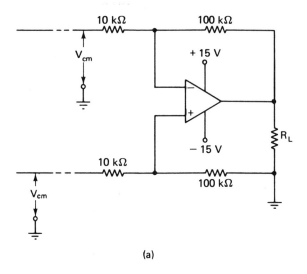

(a)

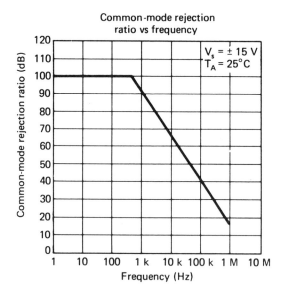

(b)

Figure 6-22 Problems 6-15, 6-16, and 6-21.

6-16. Referring to the Op Amp and its characteristics in Fig. 6-22, how much common-mode voltage can we expect across the load R_L if the input common-mode voltage $V_{cm} = 2$ mV at 100 kHz?

6-17. Referring to Fig. 6-9, if the Op Amp circuit in part a has a slew rate of 0.5 V/μs and has the input signal V_s waveform in b of the figure applied, sketch the output waveform and indicate its peak-to-peak value if the frequency of V_s is 62.5 kHz.

6-18. If a voltage-follower circuit has the square-wave input and sawtooth output waveforms of Fig. 6-23, what is the Op Amp's slew rate? The square-wave frequency $f = 1$ MHz.

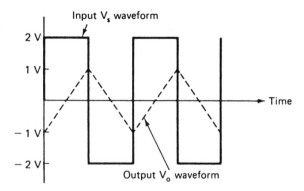

Figure 6-23   Problem 6-25.

Section 6.7

6-19. A square wave, having an extremely fast rise and fall time, is applied to the input of an Op Amp circuit. Without saturating the Op Amp, the resulting output is a trapezoidal waveform, like d of Fig. 6-9. By adjustment of the sweep time on a triggered sweep oscilloscope, we observe the leading edge of the output waveform to be as shown in Fig. 6-14. If the SWEEP TIME time control is in the CAL 0.5-μs/DIV position, what is this circuit's bandwidth?

6-20. An LF351 Op Amp is to work as a voltage follower with an unclipped output signal as large as 24 V peak to peak. What bandwidth can we expect with this output amplitude? (See Appendix E.)

Section 6.8

6-21. If the total effective input noise voltage is 5 μV rms in the circuit of Fig. 6-22, what is this circuit's output noise voltage?

6-22. If the Op Amp in the circuit of Fig. 6-18 has a specified input noise voltage of 16 nV/$\sqrt{Hz}$ and a specified input noise current of 0.01 pA/$\sqrt{Hz}$ at 1 kHz, find (a) the equivalent input noise voltage and (b) the equivalent output noise voltage at this frequency. Assume that the internal resistance of the source V_s is negligible and that the POT is set at 100 kΩ.

Section 6.4

6-23. The POT in the circuit of Fig. 6-8 is adjusted to 90 kΩ. What is the circuit's small-signal bandwidth if the Op Amp is an LF13741?

6-24. The LF351 Op Amp has a gain-bandwidth product of 4 MHz and it is pin compatible with the 741. Referring to the circuit prescribed in the previous problem, what is its small-signal bandwidth if the LF 13741 is replaced with an LF351?

Section 6.5

6-25. A voltage-follower circuit has the square-wave input shown in Fig. 6-23 at 62.5 kHz. Sketch its output waveform if the Op Amp is an LF351 that has a slew rate of 13 V/μs.

6-26. The circuit of Fig. 6-24 uses a 709 Op Amp. What is its bandwidth and its gain-bandwidth product? (See the 709's Spec in Appendix J.)

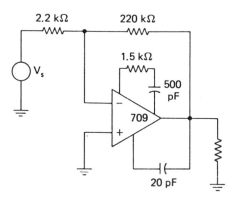

Figure 6-24 Problem 6-26.

7

PRACTICAL CONSIDERATIONS

In the preceding chapters we learned what an Op Amp's characteristics should be ideally and what they are in practice. In this chapter, we will see that many of the practical Op Amp characteristics, as listed on its specification sheets, are not constant and tend to drift with drifting ambient conditions. The effect of such drifting characteristics is usually most noticeable at the output of the Op Amp, and we will see methods of predicting output voltage changes vs. temperature and dc voltage source changes. In addition, we will see the characteristics of some special-purpose Op Amps and protecting techniques that are sometimes necessary in Op Amp circuits.

7.1 OFFSET VOLTAGE VS. POWER SUPPLY VOLTAGE

Because the Op Amp is capable of amplifying dc voltages, it is inherently sensitive to changes in its own dc supply voltages: the $+V$ and $-V$ sources. With practical Op Amps, if the dc supply voltages change due to poor regulation, the dc offset voltages will change too. Similarly, if the supply voltages are poorly filtered and vary at some ripple frequency, the offset voltages in the Op Amp will vary at the same frequency.

The sensitivity of an Op Amp to variations in the supply voltages is usually specified in a variety of somewhat equivalent terms such as the *power supply rejection ratio*, the *supply voltage rejection ratio*, the *power supply sensitivity*, and the *supply voltage sensitivity*, to name a few. These are specified in decibels or microvolts per volt. For example,

$$\text{power supply rejection ratio, PSRR, in decibels} = 20 \log \frac{\Delta V}{\Delta V_{io}}. \qquad (7\text{-}1)$$

or

$$\text{power supply sensitivity, } S, \text{ in } \mu\text{V}/\text{V} = \frac{\Delta V_{io}}{\Delta V}, \qquad (7\text{-}2)$$

where ΔV is the change in the power supply voltage, and
    ΔV_{io} is the resulting change in input offset voltage.

The significance of these parameters is more apparent if we recall that the input offset voltage V_{io} is the voltage required across the differential inputs to null the output of the Op Amp with no feedback (open loop). We also learned in Chapter 4 that, with feedback, the input offset voltage sees the circuit as a noninverting amplifier (see Fig. 4-5) and will cause an output offset voltage V_{oo} if the circuit has not been nulled. This output offset is therefore the stage gain times the input offset; that is,

$$V_{oo} = A_v V_{io} \qquad (4\text{-}1)$$

or

$$V_{oo} \cong \left(\frac{R_F}{R_1} + 1 \right) V_{io}.$$

Apparently then, any input offset voltage *change* ΔV_{io}, caused by a power supply voltage *change* ΔV, will be amplified and cause an output offset voltage *change* ΔV_{oo} that is larger by the stage gain. Equation (4-1) can therefore be modified to

$$\Delta V_{oo} \cong \left(\frac{R_F}{R_1} + 1 \right) \Delta V_{io}. \qquad (7\text{-}3)$$

Changes in the input offset voltage ΔV_{io} can be determined if we rearrange Eq. (7-1) or (7-2). If the power supply rejection ratio in decibels is

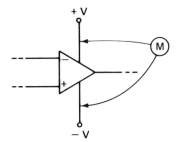

Figure 7-1   Meter M reads the total dc supply voltage.

given, then

$$\Delta V_{io} = \frac{\Delta V}{\text{antilog}(\text{PSRR (dB)}/20)}, \qquad (7\text{-}1a)$$

where the antilog of (PSRR (dB)/20) can quickly be estimated with the chart in Fig. 5-4. If a sensitivity factor S is given, then

$$\Delta V_{io} = S(\Delta V). \qquad (7\text{-}2a)$$

Since the change in supply voltage ΔV can be due to poor regulation or poor filtering, then ΔV can be the ripple voltage V_r riding on the total supply voltage. This means that, with poor filtering, $\Delta V = V_r$ and the input offset voltage varies (changes) at the ripple frequency.

Supply voltage changes ΔV are easily measured across the $+V$ and $-V$ terminals as shown in Fig. 7-1. The meter M can be a voltmeter or an oscilloscope that is capable of measuring ΔV caused by poor regulation or ΔV caused by poor filtering.

Example 7-1

Referring to Fig. 7-2, suppose the circuit is nulled when the voltage across terminals $+V$ and $-V$ measures 30 V dc and that, due to poor regulation, this dc voltage drifts with time from 28 to 32 V. Also suppose that, due to poor filtering, 2.5-mV rms ac ripple is also measured across the terminals $+V$ and $-V$. While the signal source $V_s = 0$ V, what drift occurs in the output offset voltage and how much ripple voltage can we expect across the load R_L if (a) the Op Amp's PSRR = 80 dB. Also, (b) answer these questions where the Op Amp's power supply sensitivity $S = 150 \ \mu\text{V/V}$?

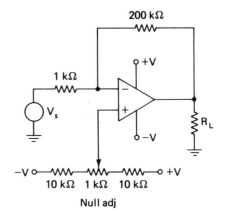

Figure 7-2   Null adj

Answer. Due to poor regulation, the dc supply can drift 2 V in either direction from 30 V; therefore, let $\Delta V = 2$ V. Due to poor filtering, $\Delta V = 2.5$ V rms.

(a) With the PSRR $= 80$ dB, which is equivalent to a voltage ratio of 10,000 (see Fig. 5-4), then due to the drift $\Delta V = 2$ V,

$$\Delta V_{io} = \frac{2 \text{ V}}{\text{antilog}(80 \text{ db}/20)} = \frac{2 \text{ V}}{10,000} = 0.2 \text{ mV}, \quad (7\text{-}1a)$$

and the resulting output offset drift is

$$\Delta V_{oo} \cong \left(\frac{R_F}{R_1} + 1\right)\Delta V_{io} = 201(0.2 \text{ mV}) = 40.2 \text{ mV}. \quad (7\text{-}3)$$

Due to the ripple $\Delta V = 2.5$ mV rms.

$$\Delta V_{io} = \frac{2.5 \text{ mV}}{10,000} = 0.25 \text{ } \mu\text{V rms}, \quad (7\text{-}1a)$$

and the output ripple voltage is

$$V_{o(\text{ripple})} \cong \left(\frac{R_F}{R_1} + 1\right)\Delta V_{io} = 201(0.25 \text{ } \mu\text{V}) = 50.25 \text{ } \mu\text{V rms}. \quad (7\text{-}3)$$

(b) With the sensitivity factor $S = 150$ μV/V, which is equivalent to 150×10^{-6}, and a dc supply drift $\Delta V = 2$ V,

$$\Delta V_{io} = S(\Delta V) = 150 \times 10^{-6}(2 \text{ V}) = 300 \text{ } \mu\text{V} = 0.3 \text{ mV}, \quad (7\text{-}2a)$$

and the resulting output offset drift

$$V_{oo} \cong 201(0.3 \text{ mV}) = 60.3 \text{ mV}. \qquad (7\text{-}3)$$

Due to the ripple $\Delta V = 2.5$ mV rms,

$$\Delta V_{io} = 150 \times 10^{-6}(2.5 \text{ mV}) = 0.375 \text{ }\mu\text{V rms}, \qquad (7\text{-}2a)$$

and this causes an output ripple voltage of

$$V_{o(\text{ripple})} \cong 201(0.375 \text{ }\mu\text{V}) \cong 75.4 \text{ }\mu\text{V rms}. \qquad (7\text{-}3)$$

Sometimes a drift in input bias current vs. dc supply voltage is specified on manufacturers' data sheets in terms of picoamperes per volt. For example, an input bias current vs. dc supply characteristic might be specified as ± 10 pA/V, which means that the input bias current might either increase or decrease by as much as 10 pA for every one-volt change in the dc supply as measured across the $+V$ and $-V$ terminals. Any drift in the input bias current will tend to cause a drift in the output offset, which is discussed further in the next section.

7.2   BIAS AND OFFSET CURRENTS VS. TEMPERATURE

The typical and maximum values of input bias current I_B and input offset current I_{io}, as specified in manufacturers' data sheets, are usually values measured at 25°C, which is about room temperature. We cannot depend on these specified currents holding to steady values if the temperature of the Op Amp drifts. Typical I_B vs. temperature and I_{io} vs. temperature curves for a general-purpose IC Op Amp are shown in Fig. 7-3. It is interesting to note that these bias and offset currents increase with decreased temperatures, which is generally the case with BJT-input IC Op Amps. The input bias current rises with higher temperatures in FET-input types of Op Amps. In either case, temperature changes cause bias and offset current changes, resulting in a drift in the output offset voltage. High-performance Op Amps are available that feature extremely low drift with temperature changes. As would be expected, their costs are higher than for the general-purpose types. The point is, we should be aware of the economical IC Op Amp's tendency to drift. If the drift is excessive for the application we have in mind, we can then consider the higher-performance types.

We concern ourselves with either the drift in input bias current ΔI_B or with the drift in the input offset current ΔI_{io}, depending on how the circuit is wired. Generally, if an Op Amp's dc resistance to ground, as seen from its

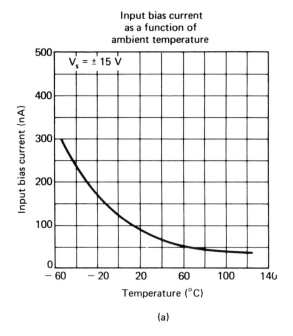

(a)

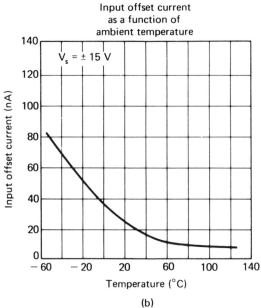

(b)

Figure 7-3   General-purpose Op Amp curves: (a) input bias current I_B vs. temperature: (b) input offset current I_{io} vs. temperature.

noninverting input 2, is negligible compared to the dc resistance to ground, seen from the inverting input 1, we can closely estimate the circuit's change or drift in output offset voltage with the equation

$$\Delta V_{oo} \cong R_F \Delta I_B,$$

(7-4)

where R_F is the feedback resistance across the inverting input 1 and the output, and

ΔI_B is the change or drift in the input bias current.

If the dc resistances to ground as seen from both the inverting and noninverting inputs are equal, the change or drift in the output offset voltage can be closely estimated with the equation

$$\Delta V_{oo} \cong R_F \Delta I_{io},$$

(7-5)

where ΔI_{io} is the change or drift in the input offset current.

For example, in Fig. 7-4a, the noninverting input 2 of the circuit is grounded; therefore, Eq. (7-4) is used to predict the drift in output offset ΔV_{oo}. Equation (7-4) also applies if R_2 is much smaller than $R_1 + R_s$ in the circuit of Fig. 7-4b and if R_s is much smaller or much larger than the resistance of R_1 and R_2 in parallel in circuit c of this figure.

On the other hand, if

$$R_2 = \frac{(R_1 + R_s)R_F}{R_1 + R_s + R_F}$$

(7-6)

in the circuit of Fig. 7-4b, and if

$$R_s = \frac{R_1 R_F}{R_1 + R_F}$$

(7-7)

in the circuit of Fig. 7-4c, the dc resistances to ground seen looking from both inputs of each circuit are equal, and Eq. (7-5) is applicable.

Since ΔI_{io} is much smaller than the ΔI_B of a given Op Amp, designing for equal dc resistances to ground reduces an Op Amp circuit's drift. All the circuits of Fig. 7-5 are low-drift (stable) designs, provided their component values are properly selected. If R_2 is selected with Eq. (7-6) in the circuit of

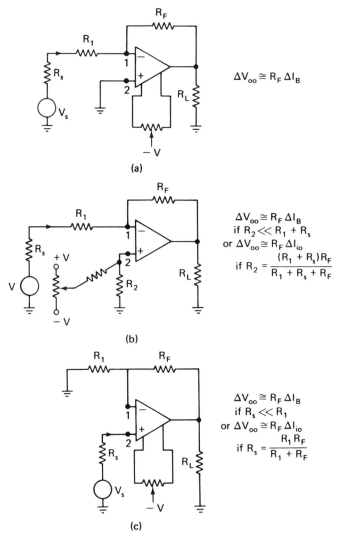

Figure 7-4   There is a change in output offset voltage ΔV_{oo} if a change in the input bias current ΔI_B occurs.

Fig. 7-5a, and if $R_a = R_1$ and $R_b = R_F$ in circuit b of the figure, and if

$$R_2 = R_s - \frac{R_1 R_F}{R_1 + R_F} \qquad (7\text{-}8)$$

in the circuit of Fig. 7-5c, the output offset drift ΔV_{oo} is closely related to the offset current drift ΔI_{io} rather than bias current drift ΔI_B which make these circuits less prone to drift with temperature changes.

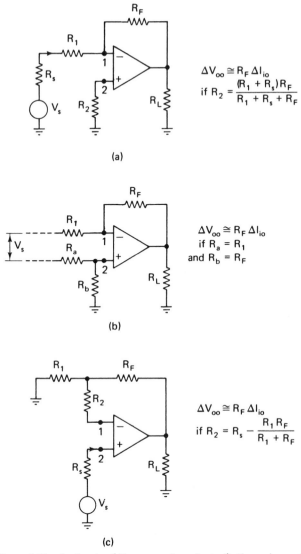

$$\Delta V_{oo} \cong R_F \Delta I_{io}$$
$$\text{if } R_2 = \frac{(R_1 + R_s)R_F}{R_1 + R_s + R_F}$$

(a)

$$\Delta V_{oo} \cong R_F \Delta I_{io}$$
$$\text{if } R_a = R_1$$
$$\text{and } R_b = R_F$$

(b)

$$\Delta V_{oo} \cong R_F \Delta I_{io}$$
$$\text{if } R_2 = R_s - \frac{R_1 R_F}{R_1 + R_F}$$

(c)

Figure 7-5   In stabilized circuits (R_2 properly selected) there is a change in the output offset voltage ΔV_{oo} if a change in the input offset current ΔI_{io} occurs.

Example 7-2

(a) The Op Amp in the circuit of Fig. 7-4b has the characteristics of Fig. 7-3 and is nulled at 25°C. What output offset is across the load R_L at 60°C if $R_s = 0 \ \Omega$, $R_1 = 10 \ \text{k}\Omega$, $R_F = 1 \ \text{M}\Omega$, and $R_2 = 100 \ \Omega$?

(b) The Op Amp in the circuit of Fig. 7-5b has the characteristics shown in Fig. 7-3 and is nulled at 25°C. What output offset is across the load R_L at 60°C if $R_a = R_1 = 10 \ k\Omega$ and $R_b = R_F = 1 \ M\Omega$?

Answers

(a) Since $R_2 \ll R_1 + R_s$, Eq. (7-4) and the curve of Fig. 7-3a apply. On this curve, $I_B \cong 80$ nA at 25°C, but it drops to 50 nA at 60°C. Thus, $\Delta I_B = 30$ nA. This causes a

$$\Delta V_{oo} \cong R_F \Delta I_B \cong 1 \ M\Omega(30 \ nA) = 30 \ mV,$$

which means that the output voltage V_o drifts from its initial 0 V at 25°C to 30 mV as the temperature rises to 60°C.

(b) In this case, since $R_a = R_1$ and $R_b = R_F$, Eq. (7-5) and the curve in Fig. 7-3b apply. On this curve, $I_{io} \cong 25$ nA at 25°C, and it decreases to about 10 nA at 60°C. Therefore, with the $\Delta T = 60 - 25 = 35$°C, the $\Delta I_{io} = 15$ nA. The resulting change in output offset voltage is

$$\Delta V_{oo} \cong R_F \Delta I_{io} \cong 1 \ M\Omega(15 \ nA) = 15 \ mV.$$

Thus as the temperature increases from 20°C to 60°C, the output offset voltage increases from its initial 0 V to either plus or minus 15 mV.

Sometimes the drift in input bias current due to temperature changes is specified in terms of picoamperes per degree Celsius over a specified temperature range. For example, some low bias current Op Amps have a specified drift of 0.001 pA/°C over the temperature range from +10°C to 70°C. Or FET-input Op Amp's data sheets might show that current simply doubles for every 10° rise in temperature in the range from −25°C to +85°C. Needless to say, specifications on drift, whether on curves or in tables, enable us to predict the maximum drift we might have in the output voltage level under known temperature variations.

7.3   INPUT OFFSET VOLTAGE VS. TEMPERATURE

The input offset voltage V_{io}, as listed on the manufacturers' data sheets, was measured at 25°C unless otherwise specified. If the temperature changes, the input offset voltage changes too. The data sheets usually specify this drift in microvolts per degree Celsius ($\mu V/°C$). We already noted, in Eq. (7-3), that a change in the input offset voltage causes a change in the output voltage of the Op Amp.

Example 7-3

In the circuit of Fig. 7-5b, $R_a = R_i = 10$ kΩ and $R_b = R_F = 1$ MΩ. It was nulled at 25°C. If the Op Amp has a specified $\Delta V_{io}/\Delta T = 15$ μV/°C, what is the output offset at 60°C as caused by the drift in V_{io}?

Answer. In this case, the temperature change

$$\Delta T = 60°C - 25°C = 35°C.$$

Therefore, the resulting change in the input offset voltage

$$\Delta V_{io} = (15 \ \mu V/°C)35°C = 525 \ \mu V.$$

By Eq. (7-3),

$$\Delta V_{oo} = \left(\frac{R_F}{R_1} + 1\right)\Delta V_{io} = \left(\frac{1 \ M\Omega}{10 \ k\Omega} + 1\right)525 \ \mu V \cong 53 \ mV.$$

7.4   OTHER TEMPERATURE-SENSITIVE PARAMETERS

A drift in the Op Amp's output voltage level is not the only possible result of temperature changes. Figure 7-6 shows a typical collection of other temperature-sensitive characteristics. The curves in Fig. 7-6a show the drift of three parameters with temperature changes on a relative scale. In this case, relative scale indicates multiplying factors that enable us to determine the changes in the three parameters if the temperature drifts away from 25°C. For example, Fig. 7-6a shows that the transient response at 70°C is 1.1 times the transient response at 25°C. On this same graph, the closed-loop BW at 70°C is shown to be 0.9 of its value at 25°C. The slew rate is shown to be constant, with temperatures above 25°C.

The transient response of an Op Amp is the time (usually in microseconds) required for its output voltage to rise from 10% to 90% of its final value under *small-signal* conditions. A typical transient response waveform is shown in Fig. 7-7. Note that the voltage level here is much smaller than that used to define the slew rate.

As shown in Fig. 7-6b, the input resistance R_i of the typical Op Amp increases with an increasing temperature. This could be a problem if we drive

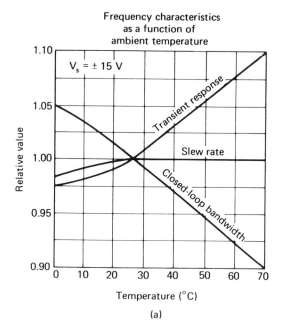

(a)

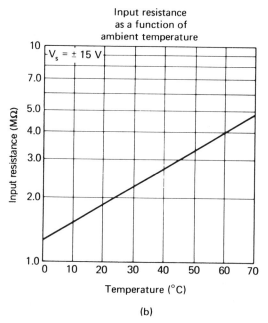

(b)

Figure 7-6   Temperature-sensitive parameters of the typical Op Amp.

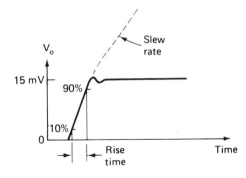

Figure 7-7   Typical transient response waveform. The rise time is the time the output voltage can change from 10% to 90% of its final level under small-signal conditions.

a noninverting amplifier with a high internal resistance signal source. A drifting input resistance, which is the load resistance on the signal source, can cause a drifting input signal amplitude.

7.5   CHANNEL SEPARATION

If a monolithic Op Amp package contains more than one amplifier (dual or quad Op Amp), a parameter called *channel separation* is specified in its data sheets. It tells us how much interaction we can expect between the amplifiers. If a signal is applied to the input of one amplifier in, say, a dual package, some signal will appear at the output of the other Op Amp even though it has no input signal applied. This interaction exists because of the close physical proximity, which causes electrical coupling, between the two Op Amps built on a single semiconductor chip.

A 747 package is shown in Fig. 7-8. It contains essentially two 741 Op Amps. Its channel separation is typically specified at a minimum of 100 dB.

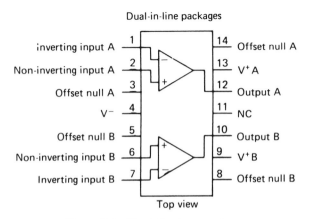

Figure 7-8   Dual Op Amp package.

This means that if one of the Op Amps is driven while the other is not, the output signal of the undriven one will be at least 100 dB below the signal output of the driven amplifier.

Example 7-3

If one of the Op Amps of a 747 is driven so that its output signal is 15 V peak to peak at 400 Hz, how much 400-Hz signal might be at the output of the other amplifier even though it is not driven?

Answer. Since the 747's channel separation is at least 100 dB, which is equivalent to a ratio of 10^5, the output of the undriven amplifier will not be more than

$$\frac{15 \text{ V}}{10^5} = 150 \text{ } \mu\text{V peak to peak.}$$

7.6 CLEANING PC BOARDS AND GUARDING INPUT TERMINALS

As mentioned previously, some applications require Op Amps with low bias currents. Certain types of high-beta transistor-input Op Amps, and FET-input Op Amps are made for such applications. Use of such low-bias-current Op Amps presents new problems, however. When the low-bias-current Op Amp is used on a printed circuit (PC) board, the circuit tends to behave erratically and its dc output may drift with temperature changes, especially in higher temperature ranges, if certain precautions are not taken.

A drift problem can be caused by PC board leakage currents. These currents can easily exceed the bias currents of the Op Amp, especially if it is a low-bias-current type. Such a leakage current might flow between either rail voltage to one of the Op Amp inputs. This current through the board leakage resistance can cause an erratic voltage at the affected input, which, in turn, causes an unstable (drifting) output voltage. To reduce the possibility of such drift, the printed circuit board must be thoroughly cleaned with an appropriate solvent to remove all solder flux. The board can then be coated with silicone rubber or epoxy to keep its surface clean. Input *guards* help too. Figure 7-9 shows a PC board layout with a guard around the inputs, along with its schematic representation and proper termination in common circuits.

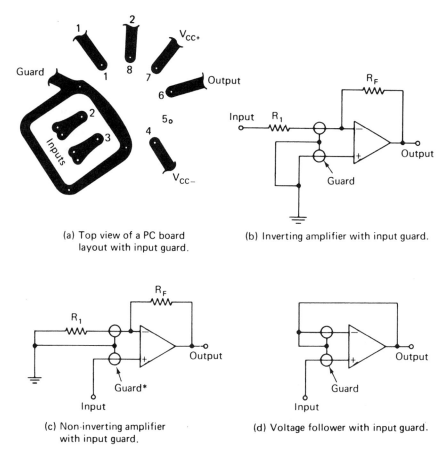

(a) Top view of a PC board layout with input guard.

(b) Inverting amplifier with input guard.

(c) Non-inverting amplifier with input guard.

(d) Voltage follower with input guard.

Figure 7-9 (a) Top view of a PC board layout with input guard, (b) inverting amplifier with input guard, (c) noninverting amplifier with input guard, (d) voltage follower with input guard.

7.7 PROTECTING TECHNIQUES

It takes on the order of just a few millivolts across an Op Amp's inverting and noninverting inputs to drive its output into saturation. Large differential inputs can ruin it. If large input peaks or transients are expected across the inputs, the Op Amp can be protected as shown in Fig. 7-10. The diodes remain nonconducting and therefore do not affect the input signals as long as these signals are small, as they normally should be. Large input signals, however, drive the diodes into conduction. Thus the differential input voltages are limited to a few hundred millivolts—the forward voltage drop of each diode.

Figure 7-10   Input breakdown protection.

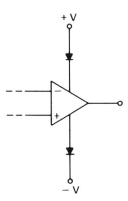

Figure 7-11   dc supply voltage reversal protection.

Op Amps can be ruined if the dc supply voltages are connected in the wrong polarities. Diodes in series with the dc supply leads, such as in Fig. 7-11, will protect the Op Amp from such an accident.

If unregulated or poorly regulated dc supplies are used, the Op Amp can be protected from excessive supply voltage with a zener (regulator diode) as shown in Fig. 7-12. Thus if the Op Amp's maximum dc supply voltages are specified at ± 17 V, a zener with a breakdown voltage of 34 V or less would be used.

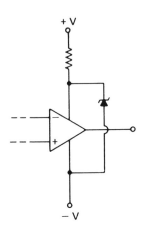

Figure 7-12   dc supply overvoltage protection.

REVIEW QUESTIONS 7

7-1. If a hypothetical Op Amp has an infinite power supply rejection ratio, how stringent should its power supplies' regulation and filtering be?

7-2. What are some causes of drift in the output voltage level of an Op Amp with temperature changes?

7-3. From which would we expect more drift: an Op Amp with relatively large or a relatively low specified input bias current? Why?

7-4. In the circuit of Fig. 7-4a, both the input bias current and the input offset current change with temperature. Which of these enables us to predict the output voltage drift? Why?

7-5. In the circuit of Fig. 7-5a, if we know the Op Amp's change in input bias current and its change in input offset current with temperature, which of these specifications enables us to predict the output voltage drift? Why?

7-6. Name five Op Amp parameters that drift with temperature changes.

7-7. Compare the parameters of Op Amp types LM741C and LF351A. Which would you expect to have more drift? Why? (See Appendix E.)

7-8. What is the purpose of resistor R_2 in circuit c of Fig. 7-5?

7-9. What PC board precautions should be followed when Op Amps having extremely low bias currents are used?

7-10. If the ambient temperature is subject to considerable changes, which is more stable with a given type of general-purpose Op Amp, the circuit in Fig. 7-4a or the circuit in Fig. 7-5a? Why?

PROBLEMS 7

Section 7.1

7-1. In the circuit of Fig. 7-2, suppose that we null the output when the voltage across the $+V$ and $-V$ supply terminals measures 24 V. This voltage then drifts from 20 to 28 V. If the Op Amp's PSRR = 100 dB, what changes can we expect in its (a) input offset voltage, and (b) output voltage level?

7-2. In the circuit of Fig. 7-5b, what drift in input offset voltage and output voltage can we expect if $R_1 = R_a = 10$ kΩ, $R_F = R_b = 1$ MΩ, the power supply sensitivity $S = 100$ μV/V, and the voltage across the $+V$ and $-V$ supply terminals drifts in the range from 20 to 25 V?

7-3. In the circuit described in Problem 7-1, how much 60-Hz noise can we expect across the load if there is a 4-mV rms, 60-Hz voltage across the $+V$ and $-V$ pins?

7-4. In the circuit described in Problem 7-2, how much 60-Hz noise can we expect across the load if a 6-m V rms, 60-Hz voltage is measured across the $+V$ and $-V$ terminals?

7-5. If an Op Amp has a PSRR of 100 dB, what is its power supply sensitivity S in $\mu V/V$?

7-6. If an Op Amp has a power supply sensitivity $S = 200~\mu V/V$, what is its PSRR in decibels?

Section 7.2

7-7. In the circuit of Fig. 7-4a, suppose that $R_s = 1000~\Omega$, $R_1 = 20~K\Omega$, $R_F = 10~M\Omega$, and the circuit is nulled at 25°C. What dc output could we expect at 65°C if the Op Amp is an FET type whose bias current is 30 pA at 25°C and whose bias current doubles for every 10°C rise in temperature?

7-8. In the circuit of Fig. 7-4a, suppose that $R_s = 100~\Omega$, $R_1 = 1~k\Omega$, $R_F = 20~k\Omega$, and the circuit is nulled at 20°C. What dc output voltage can we expect at 0°C if the Op Amp has the characteristics shown in Fig. 7-3?

7-9. If in the circuit of Fig. 7-5a, $R_s = 200~\Omega$, $R_1 = 2~k\Omega$, $R_2 = 2~k\Omega$, $R_F = 20~k\Omega$, and the dc output is 500 mV at 20°C, what is the approximate maximum possible output offset voltage at (a) 0°C and at (b) 60°C? Assume that the Op Amp has characteristics as shown in Fig. 7-3.

7-10. Assume that in the circuit of Fig. 7-4b, $R_1 = 1~k\Omega$, $R_2 = 1~k\Omega$, $R_s = 100~\Omega$, and $R_F = 10~k\Omega$. If the Op Amp has the characteristics shown in Fig. 7-3, and if the circuit is nulled at 20°C, what dc output can we expect at temperatures (a) $-20°C$ and (b) 100°C?

Section 7.3

7-11. If the FET-input Op Amp in the circuit described in Problem 7-7 has a $\Delta V_{io}/\Delta T = 20~\mu V/°C$, (a) what drift in output offset could we expect at 65°C as caused by the drift in V_{io}? (b) What is the drift in output offset as caused by the combined effects of the drifts in bias current and input offset voltage?

7-12. If in the circuit of Fig. 7-4a, $R_s = 0$, $R_1 = 2.2\ \text{k}\Omega$, $R_F = 820\ \text{k}\Omega$, and the Op Amp is an LF351 type, what drift in the output voltage can we expect if the temperature can vary from 25 to 65°C? (See Appendix E for specs on the LF351.)

Section 7.4

7-13. If an Op Amp has the characteristics shown in Fig. 7-6a and has a 100-ns transient response at 25°C, what is its transient response at (a) 0°C and at (b) 70°C?

7-14. If an Op Amp has the characteristics shown in Fig. 7-6a and has a closed-loop bandwidth of 40,000 Hz at 25°C, what are its bandwidths at temperatures (a) 0°C and (b) 70°C?

Section 7.3

7-15. If $R_1 = R_a = 1.1\ \text{k}\Omega$ and $R_F = R_b = 680\ \text{k}\Omega$ in the circuit of Fig. 7-5b, what is the output offset voltage V_{oo} as caused by bias current (a) if the Op Amp is a 741 type, and (b) if the Op Amp is an LF351 type? Note: The LF351 is a FET-input Op Amp that is pin compatible with the 741. (See Appendix E.)

7-16. Find the value of R_2 in the circuit of Fig. 7-5c if $R_1 = 1\ \text{k}\Omega$, $R_F = 10\ \text{k}\Omega$ and the internal resistance of the signal source $R_s = 40\ \text{k}\Omega$. With your value of R_2 in this circuit, what is its closed-loop voltage gain?

8

ANALOG APPLICATIONS OF OP AMPS

Electronic circuits are classified as being either analog or digital. Op Amps are designed to work as analog devices, though as we will see later they are used in digital circuits too. An analog circuit has a continuously variable output voltage or current that is a function of an input voltage or current. All amplifiers described in the previous chapters are analog applications of Op Amps. Amplifiers, more specifically, fall into a category called *linear circuits*, which is a subset of analog circuits. Generally, a linear circuit has an output voltage or current that is proportional to (a multiple of) an input voltage or current. Needless to say, an Op Amp can work in an analog mode only if its output voltage and frequency are within its capabilities. An overview of previously described circuits is given here, along with discussions of other common analog applications of Op Amps.

8.1 OP AMPS AS AC AMPLIFIERS

The Op Amp applications discussed till now were mainly dc amplifiers. In other words, their output voltages change in respose to change in dc inputs. All the circuits in Fig 8-1 were discussed previously and are dc type amplifiers. Of course, they respond linearly to ac input signals too, provided the frequencies are not too high. In some applications, the circuit designer needs the ac response characteristics of the Op Amp but does not need or want its dc response capability. A multistage audio amplifying system is an example. Even a small dc input offset voltage in the first stage can be amplified to a value large enough to saturate a following Op Amp stage if the stages are directly coupled. Capacitive coupling, as shown in Fig. 8-2, between stages is

142

a simple way of eliminating dc level amplification from stage to stage. The signal source V_s could be a preceding amplifier with a dc component voltage at its output. The coupling capacitor C effectively blocks this dc, preventing it from affecting the next stage. The resistor R_2 *must* be used because it provides a dc path between the noninverting input and ground. Without R_2, the coupling capacitor C would become charged by the bias current I_{B_2} and would cause a dc input voltage on the noninverting input, causing a dc output that runs into the rail.

Since the circuit in Fig. 8-2 is basically a noninverting amplifier, its closed-loop gain is

$$A_v \cong \frac{R_F}{R_1} + 1 \qquad (3\text{-}6)$$

at frequencies within its bandwidth. The low-frequency limit of the bandwidth is determined largely by the input resistance and the size of the coupling capacitor C.*

An inverting ac amplifier is shown in Fig 8-3. If the reactance of C is negligible, its gain is

$$A_v \cong -\frac{R_F}{R_1} \qquad (3\text{-}4)$$

at frequencies within its bandwidth. If high gain or a large feedback resistor R_F is used, a resistor R_2 equal to R_F is used between the noninverting input and ground to reduce the output offset and drift caused by bias current.

With negligible reactances of the coupling capacitors C, the signal source V_s sees R_2 as the stage input resistance in the circuit of Fig 8-2, and it sees R_1 as the input resistance in the circuit of Fig 8-3. These resistances are relatively low, and in cases where the ac signal source V_s must work into a very large stage input resistance, the amplifier's input resistance can be *bootstrapped*, as shown in Fig. 8-4. This circuit is a voltage follower. Its output signal V_o is applied to the bottom of R_1 via capacitor C_2, while the input signal V_s is applied to the top of R_1 through the coupling capacitor C_1. Therefore, the signal potential difference across R_1 is $V_s - V_o$. But since $V_s \cong V_o$ in a voltage follower, their difference is extremely small, which means that the signal current through R_1 is also extremely small. Since the signal current in R_1 is also the current drain from the signal source V_s, the impedance seen by the signal source V_s is extremely large. Resistor R_F is selected to be equal to the sum of R_1 and R_2. This reduces the drift of the output offset due to drifting bias current.

*See Appendix K for a method of selecting the proper coupling capacitor values.

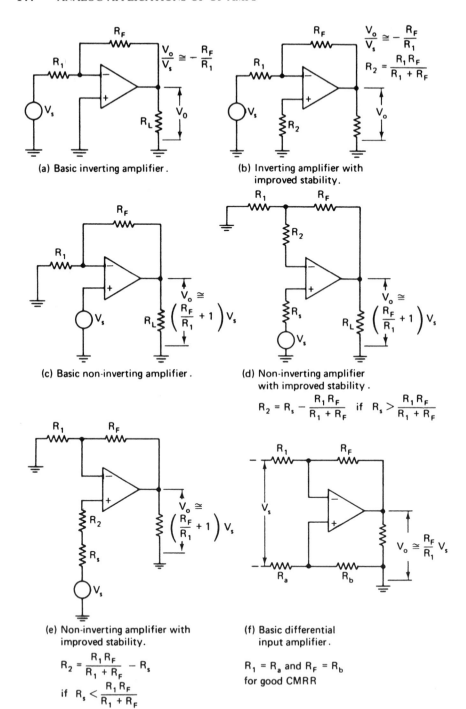

Figure 8-1 Common linear Op Amp applications.

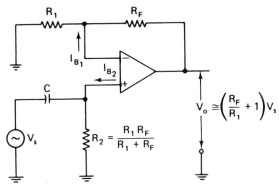

Figure 8-2   An ac noninverting amplifier; I_{B_1} and I_{B_2} are dc bias currents.

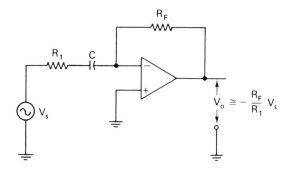

Figure 8-3   ac inverting amplifier.

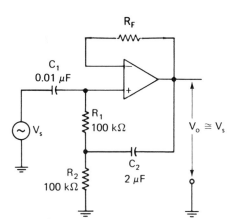

Figure 8-4   ac voltage follower with input impedance bootstrapped (increased).

8.2 SUMMING AND AVERAGING CIRCUITS

This circuit in Fig. 8-5 is the basic inverting summing and averaging circuit. The value and polarity of the output voltage V_o is determined by the sum of the input voltages V_1 through V_n and by the values of the externally wired resistors. Since the noninverting input is at ground potential, the inverting input and the right sides of input resistors R_a through R_n are *virtually* grounded too. Thus, the currents in the input resistors can be shown, by Ohm's law, as

$$I_a \cong \frac{V_a}{R_a},$$

$$I_b \cong \frac{V_b}{R_b},$$

$$I_c \cong \frac{V_c}{R_c},$$

$$I_n \cong \frac{V_n}{R_n}.$$

Similarly, the current in R_F is

$$I_F \cong \frac{-V_o}{R_F},$$

where the negative sign indicates that V_o is out of phase with the net voltage at the inverting input. Since the sum of the input currents is equal to the

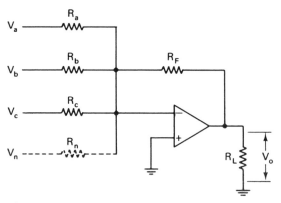

Figure 8-5   Basic inverting summing or averaging circuit.

feedback current, assuming that the Op Amp is ideal, we can show that

$$I_F \cong I_a + I_b + I_c + \cdots + I_n$$

or

$$\frac{-V_o}{R_F} \cong \frac{V_a}{R_a} + \frac{V_b}{R_b} + \frac{V_c}{R_c} + \cdots + \frac{V_n}{R_n}.$$

Multiplying both sides of the previous equation by $-R_F$ shows that, generally, the output voltage is

$$V_o \cong -R_F \left(\frac{V_a}{R_a} + \frac{V_b}{R_b} + \frac{V_c}{R_c} + \cdots + \frac{V_n}{R_n} \right). \tag{8-1}$$

If all input resistors are equal, say

$$R_a = R_b = R_c = R_n = R,$$

Eq. (8-1) simplifies to

$$V_o \cong -\frac{R_F}{R}(V_a + V_b + V_c + \cdots + V_n). \tag{8-2}$$

If $R_F = R$, this equation simplifies further to

$$V_o \cong -(V_a + V_b + V_c + \cdots V_n). \tag{8-3}$$

This last equation shows that the circuit in Fig. 8-5 can be used to find the sum of any number of input voltages.

This circuit can also average the input voltages by using a ratio of R_F/R that is equal to the reciprocal of the number of voltages being averaged. The ratio R_F/R is selected so that the sum of the input voltages is divided by the number of input voltages applied.

Of course, the above equations apply assuming that the Op Amp is properly nulled; that is, $V_o = 0$ V if all inputs are 0 V. The effect that input bias current has on the output offset and stability can be reduced if we use a resistor R_2 between the noninverting input and ground; see Fig. 8.6, where

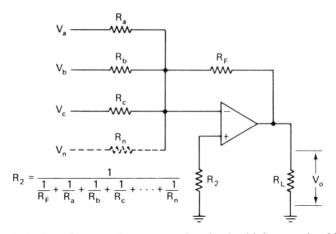

Figure 8-6   Inverting summing or averaging circuit with improved stability.

its value is

$$R_2 = \frac{1}{1/R_F + 1/R_a + 1/R_b + 1/R_c + \cdots + 1/R_n}. \tag{8-4}$$

The input voltage sources and resistors can be connected to the noninverting input of an Op Amp as shown in Fig. 8-7a. These sources and resistors can be replaced with a Norton's equivalent current, as shown in Fig. 8-7b, where

$$I_{eq} = \frac{V_a}{R_a} + \frac{V_b}{R_b} + \frac{V_c}{R_c} + \cdots + \frac{V_n}{R_n} \tag{8-5}$$

and

$$R_{eq} = \frac{1}{1/R_a + 1/R_b + 1/R_c + \cdots + 1/R_n}. \tag{8-6}$$

Since the resistance looking into the noninverting input is very large, practi-

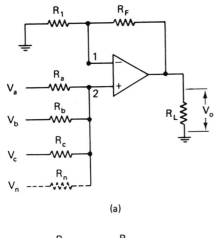

(a)

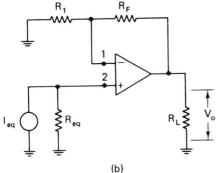

(b)

Figure 8-7 Noninverting averaging circuit.

cally all of the current I_{eq} is forced through R_{eq}, resulting in a voltage at the noninverting input 2; that is,

$$V_2 = R_{eq} I_{eq}.$$ (8-7)

If all of the input resistors are equal, voltage V_2 is the average of the input voltages. The resulting output voltage is

$$V_o \cong \left(\frac{R_F}{R_1} + 1\right) V_2.$$ (8-8)

Example 8-1

Determine the value and polarity of the output voltage in each of the circuits in Fig. 8-8.

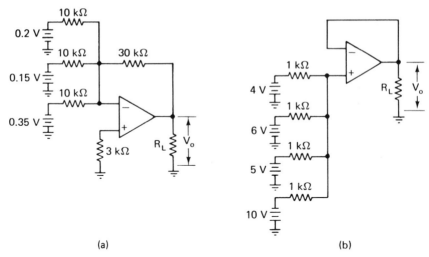

(a)   (b)

Figure 8-8

Answer. The circuit shown in Fig. 8-8a is an inverting summing amplifier. Since all of its input resistors are equal, the output voltage can be determined with Eq. (8-2); that is,

$$V_o \cong -\frac{30\ k\Omega}{10\ k\Omega}(0.2\ V - 0.15\ V - 0.35\ V)$$

$$= -30(-0.3\ V) = +0.9\ V.$$

The circuit in Fig 8.8b is a noninverting type whose input voltages and resistances are replaced with a Norton's equivalent circuit where

$$I_{eq} = \frac{-4\ V}{1\ k\Omega} + \frac{6\ V}{1\ k\Omega} + \frac{5\ V}{1\ k\Omega} + \frac{-10\ V}{1\ k\Omega}$$

$$= -4\ mA + 6\ mA + 5\ mA - 10\ mA = -3\ mA \qquad (8\text{-}5)$$

and

$$R_{eq} = \frac{1\ k\Omega}{4} = 250\ \Omega. \qquad (8\text{-}6)$$

Therefore, the voltage at the noninverting input

$$V_2 = 250\ \Omega(-3\ mA) = -0.75\ V, \qquad (8\text{-}7)$$

which is the average of the input voltages. Since this Op Amp is connected to work as a voltage follower, its output voltage V_o is the same as the input voltage, or -0.75 V in this case.

8.3   THE OP AMP AS AN INTEGRATOR

The Op Amp in Fig. 8-9 is connected to work as an integrator. Its output-voltage waveform is the negative integral of the input-voltage waveform for properly selected values of R and C. Its operation can be analyzed as follows: Assuming that the Op Amp is ideal, the inverting input 1 and the right side of the input resistor R are at ground potential because the noninverting input is grounded. Therefore, the applied voltage V_s appears across R and the current in this resistor is

$$I \cong \frac{V_s}{R}. \tag{8-9}$$

Because of the very large resistance looking into the inverting input, practically all of this current is forced through the capacitor C, which changes the voltage across it. Generally, the current through and the voltage across a capacitor are related by the equation

$$i_c = C\frac{dv_c}{dt} \tag{8-10a}$$

or

$$i_c = C\frac{\Delta v_c}{\Delta t}, \tag{8-10b}$$

where Δv_c is the change in voltage across the capacitor and Δt is the

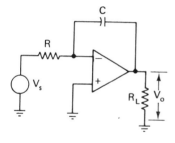

Figure 8-9   Op Amp wired to work as an integrator.

corresponding change in time. Since the left side of the capacitor C is virtually grounded, the voltage across C *is* the output voltage V_o. And since the current through C is also the current though R, the above equation can be shown as

$$I = C\frac{\Delta V_o}{\Delta t}. \qquad (8\text{-}10c)$$

And by rearranging, this equation becomes

$$\Delta V_o = \frac{I}{C}\Delta t. \qquad (8\text{-}10d)$$

This equation enables us to determine the *change* in the output voltage V_o in response to a step voltage at the input. Square and rectangular waves have step voltages on their leading and trailing edges and are very common in Op Amp applications, especially when receiving signals from the digital world.

As an example, make $R = 1\ M\Omega$ and $C = 1\ \mu F$ in the circuit of Fig. 8-9. Suppose further that the voltage V_s is the square wave shown in Fig. 8-10a.

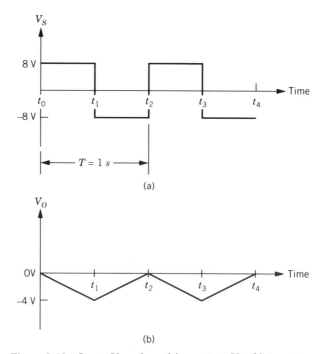

(a)

(b)

Figure 8-10 Input V_s and resulting output V_o of integrator.

Since $V_s = 8$ V in the time frame t_0 to t_1, the current during this time is

$$I = \frac{8 \text{ V}}{1 \text{ M}\Omega} = 8 \text{ }\mu\text{A}.$$

By substituting this into Eq. (8-10d), we can show that the resulting change at the output voltage is

$$\Delta V_o = \frac{8 \text{ }\mu\text{A}}{1 \text{ }\mu\text{F}} (0.5 \text{ s}) = 4 \text{ V}. \tag{8-10d}$$

When V_s is positive, the conventional current flows from left to right and the capacitor's voltage changes in the negative direction. In this case, if initially $V_o = 0$ V, the output voltage ramps negatively to -4 V in the t_0 to t_1 time frame, as shown in Fig. 8-10b. In the next time frame, t_1 to t_2, the step input voltage is -8 V and the resulting current is -8 μA. This current flows from right to left and causes a positive change by 4 V in this next 0.5 s. Note in Fig. 8-10b that V_o ramps positively at times when V_s is -8 V.

Example 8-2

The waveform of Fig. 8-11a is applied to circuit b of this figure. Sketch the resulting output.

Answer. First we determine that the current

$$I = \frac{-10 \text{ V}}{1 \text{ M}\Omega} = -10 \text{ }\mu\text{A}. \tag{8-10c}$$

This current flows for 0.1-s intervals causing a change in output voltage

$$\Delta V_o = \frac{-10 \text{ }\mu\text{A}}{1 \text{ }\mu\text{F}} (0.1 \text{ s}) = 1 \text{ V} \tag{8-10d}$$

in each interval. Assuming that the capacitor is initially uncharged ($V_o = 0$ V), the resulting output waveform is as shown in Fig. 8-11c.

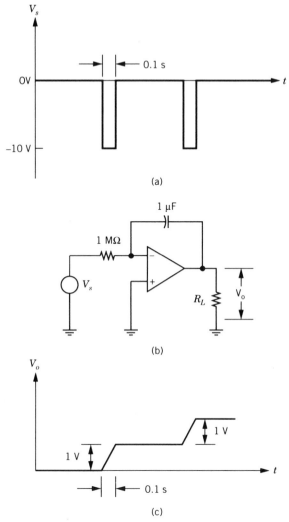

Figure 8-11   Waveforms and circuit for Example 8-2.

The circuit of Fig. 8-12 shows an integrator modified to work as a tachometer. Resistor R_D provides a discharge path. The input pulses are from the engine's ignition system. The capacitor and output voltages rise with each input pulse. The discharge path reduces these voltages between pulses. Thus V_o increases or decreases when the distance between pulses decreases or increases, respectively, and the output meter M can be calibrated in rpm of the engine.

The circuit of Fig. 8-10b is called an integrator because it performs mathematical integration. This is explained in Appendix L, which also

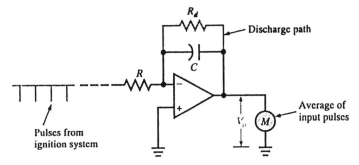

Figure 8-12   Analog tachometer circuit.

contains examples of output vs. input waveforms. Differentiation with an Op Amp is described in Appendix M.

8.4 THREE-TERMINAL REGULATORS

Manufacturers of semiconductor components make an impressive variety of inexpensive IC regulators that are easy to use. The simplest—three-terminal regulators—are able to provide fixed regulated output voltages. Three-terminal regulators are used often to supply power to Op Amp circuits. Also, since they are often part of variable regulated supplies that include Op Amps, three-terminal regulators are described here at length.

As shown in Fig. 8-13 three-terminal regulators are connected between the unregulated dc voltage source V_{in} and the load that requires the regulated voltage V_{out}. A partial list of common three-terminal regulators, given in Appendix R, shows some typical regulated output voltages: 6 V, 8 V, 12 V, 15 V, etc. Many types of three-terminal regulators can deliver over 1 A to their load while regulating V_{out}.

In Fig. 8-13a the LM340K-15 is a positive 15-V regulator.* If the regulator is located more than 5 cm (about 2 inches) from the filter capacitor C, use of an input capacitor C_{in}, which helps to maintain stability, is recommended. If needed, C_{in} should be at least a 0.22-μF ceramic disk, a 2-μF solid tantalum, or a 25-μF electrolytic. An output capacitor C_{out} serves to limit high-frequency noise and improve transient response. Good transient response on a regulator means that its output voltage V_{out} remains constant (regulated) even though there are rapid changes in the load current. Three-terminal regulators are often used on individual PC cards and in subsystems to reduce *cross-talk* and PS *distribution noise* that occurs in systems with a common PS.

In Fig. 8-13b, the LM320K-15 is a negative 15-V regulator. As with the positive regulator, C_{in} is added if there is some distance between the filter

*See Appendix S for an interpretation of the alphanumerically coded numbers.

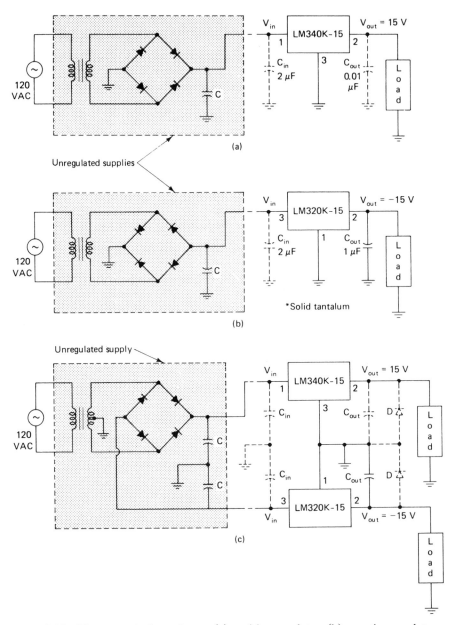

Figure 8-13   Three-terminal regulators: (a) positive regulator, (b) negative regulator, and (c) dual supply.

capacitor C and the regulator. The chip (IC) manufacturers recommend that C_{out} of 1 μF or larger solid tantalum, or a 25 μF or larger aluminum electrolytic, be used to assure stability.

As shown in Fig. 8-13c, positive and negative regulators can work together in a dual supply that has both 15 V and -15 V outputs. The diodes D at the outputs are recommended if this type of dual supply is to provide power to a common load, such as Op Amps.

The unregulated voltage V_{in} typically varies with changes in the ac line voltage and load current. The minimum value of V_{in} must always be larger than the regulated output V_{out} by the *dropout* voltage. Dropout voltages are specified by the regulator manufacturers and are typically about 2 V. Also specified are quiescent currents, I_Q. For a given regulator, the quiescent current is the difference in the input and load currents. It flows in pin 3 of the LM340K-15 and in pin 1 of the LM320K-15 in the circuits of Fig. 8-13.

Example 8-3

If in the circuit of Fig. 8-13a, V_{in} varies from 20 to 28 V and the load draws 0–150 mA, determine (a) the minimum and maximum voltages across pins 1 and 2; (b) the dropout voltage and quiescent current of the regulator (refer to the table of Appendix R); and (c) the minimum and maximum currents that this regulator draws from the bridge rectifier.

Answers

(a) According to Appendix R, the output voltage at pin 2 is 15 V. Since the unregulated voltage at pin 1 varies from 20 to 28 V, the voltage across these pins can be as low as 20 V $-$ 15 V $=$ 5 V or as high as 28 V $-$ 15 V $=$ 13 V.

(b) Appendix R shows that the LM340K-15 has a specified minimum input voltage of 17 V and an output of 15 V. Their difference, 2 V, is the dropout voltage. If V_{in} becomes less than 17 V, the regulator drops out of regulation.

(c) When the load draws 0 A, the regulator draws 6 mA; the quiescent current. When the load draws 150 mA, the regulator requires 156 mA.

8.5 REGULATOR POWER DISSIPATION

Three-terminal regulators dissipate power while performing their regulating function. The approximate value of maximum power dissipation $P_{D(max)}$ can be determined by multiplying the maximum voltage across the regulator's input and output terminals by the maximum load current—that is,

$$P_{D(\text{max})} \cong \left(V_{\text{in(max)}} - V_{\text{out}}\right) I_{L(\text{max})}. \tag{8-11}$$

Power dissipation P_D causes heat at the junctions of the transistors within the regulator. This heat flows from the junctions to the case (package) of the regulator. The heat then flows from the case to the surrounding air or to a heat sink. If used, a heat sink, in turn, transfers the heat to the air. Needless to say, $P_{D(\text{max})}$ should never exceed the power rating of the regulator.

8.6   HEAT SINKS

Regulators are available in a variety of package types (see Appendices R and S). The package of a regulator is called its *case*. Often the case serves as the ground lead (pin) on positive regulators and as the input lead on negative regulators. When required to dissipate about 1 W or more, regulators are housed in packages that are made to be mounted on a heat sink. If the case is not the ground lead, an electrically insulating washer is used between the case and the sink. Heat-conducting silicone grease is often used between the case and sink, or on both sides of the washer.

Heat generated in a regulator must first flow through its *junction-to-case* thermal resistance θ_{JC}. From the case, the heat flow continues through a *case-to-sink* thermal resistance θ_{CS} (see Table 8-1). Finally, the heat flow is through the *sink-to-air* thermal resistance θ_{SA}.

The amount of power, in the form of heat, that flows from a regulator's junctions to the surrounding air can be determined by an equation similar to Ohm's law:

$$P_D = \frac{T_J - T_A}{\theta_{JC} + \theta_{CS} + \theta_{SA}}, \tag{8-12}$$

TABLE 8-1. Thermal Resistance of Insulating Washers

Material	Dry	With Silicone
Mica	0.8	0.4
Teflon	1.45	0.8
Anodized aluminum	0.4	0.35
No insulator	0.2	0.1

where P_D is the dissipated power (heat) flow from the junctions to air in watts.

T_J is the temperature of the transistors' junctions within the regulator in degree Celsius (°C).

T_A is the ambient (surrounding air) temperature in °C.

θ_{JC} is the junction-to-case thermal resistance, in °C/W, which is specified by the regulator manufacturer.

θ_{CS} is the case-to-sink thermal resistance, in °C/W, which is obtained from Table 8-1 or its equivalent.

θ_{SA} is the sink-to-air thermal resistance, in °C/W, which is specified by the manufacturer of the heat sink.

The term $T_J - T_A$ is the difference in the junction and ambient air temperatures. As larger voltages cause larger currents in electrical circuits, so likewise do larger temperature differences cause larger power (heat) flow. Also, as smaller circuit resistance admits more current, then similarly, smaller thermal resistance allows for increased heat flow. If we are to select the smallest heat sink necessary, we need to know the largest θ_{SA} that will be adequate; physically larger heat sinks have smaller values of θ_{SA}. We can start by rearranging Eq. (8-12) to the following:

$$\theta_{SA} = \frac{T_J - T_A}{P_D} - \theta_{JC} - \theta_{CS}. \qquad (8\text{-}13)$$

When using the above equation, make P_D the maximum power that the regulator is required to dissipate and T_J the maximum recommended temperature which usually is provided on the spec sheets.

8.7 VARIABLE REGULATED POWER SUPPLY

Some applications require variable regulated power supply voltages. A three-terminal regulator can be wired as in Fig. 8-14 for such applications. In this circuit, the LM 340K-5 maintains a well regulated 5 V across its pins 2 and 3 and, therefore, across resistor R_1 too. The output voltage V_{out} is the sum of the voltages across R_1 and R_2. If we change the resistance of R_2, the voltage drop across it changes and this changes the output voltage V_{out}. If, say, R_2 is adjusted to 0 Ω, pin 3 is pulled to ground potential and $V_{out} = 5$ V. Increasing R_2 will cause V_{out} to increase to a maximum value that is V_{in} minus the dropout voltage.

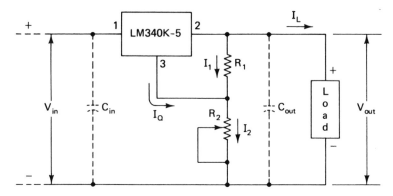

Figure 8-14   Three-terminal regulator wired to work as a variable dc voltage source.

Example 8-4

In the circuit of Fig. 8-14, V_{in} varies from 15 to 20 V, $I_Q = 3$ mA, $R_1 = 2$ kΩ, and R_2 is a 1-kΩ POT. What is V_{out}

(a) when the POT resistance is 1 kΩ, and
(b) when the POT resistance is 500 Ω?

Answer.   Since the voltage across pins 2 and 3 is regulated at 5 V, the current in R_1 is

$$I_1 = \frac{5 \text{ V}}{R_1} = \frac{5 \text{ V}}{2 \text{ k}\Omega} = 2.5 \text{ mA}.$$

This adds to I_Q to cause 5.5 mA in R_2. When (a) $R_2 = 1$ kΩ, the resulting drop across it is 1 kΩ(5.5 mA) or 5.5 V. The sum of the voltages across R_1 and R_2, therefore, is

$$V_{out} = 5 \text{ V} + 5.5 \text{ V} = 10.5 \text{ V}.$$

When (b) $R_2 = 0.5$ kΩ, its voltage drop is 0.5 kΩ(5.5 mA) or 2.75 V. In this case,

$$V_{out} = 5 \text{ V} + 2.75 \text{ V} = 7.75 \text{ V}.$$

Example 8-5

Referring to the regulator circuit described in the previous example, what maximum power will the regulator chip dissipate if the load draws up to 200 mA?

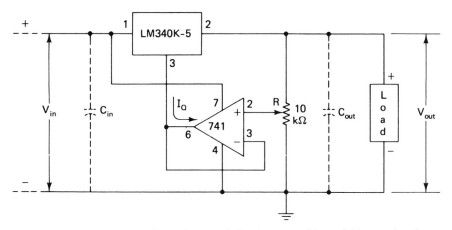

Figure 8-15 Three-terminal regulator and Op Amp provide variable regulated output voltage V_{out}.

Answer. The maximum voltage across the regulator (pins 1 and 2) occurs when V_{in} is maximum while V_{out} is adjusted to its minimum value. In this case, this maximum voltage is 20 V − 7.75 V = 12.25 V. Since I_Q and I_1 are negligible compared to the 200-mA maximum load current, the maximum regulator power

$$P \cong 12.25 \text{ V}(200 \text{ mA}) = 2.45 \text{ W}.$$

The circuit of Fig. 8-14 is simple and can provide a moderately well regulated adjustable output voltage while using a fixed voltage regulator. It has its shortcomings, however. The quiescent current I_Q of the regulator is not constant—it varies with load. Variations in load, therefore, cause variations in I_Q, which in turn cause variations in voltage across the potentiometer R_2. Since the sum of the drops across R_1 and R_2 is equal to the output voltage V_{out}, this circuit cannot take advantage of the regulator's ability to regulate.

Compare the circuit of Fig. 8-15 with that of Fig. 8-14. Note that an Op Amp is used here to *sink* quiescent current I_Q. In this case, I_Q "sees" the extremely low output resistance ($R_{o(eff)}$) of the Op Amp on its way to ground instead of R_2. Variations in I_Q, therefore, have negligible influence on voltages in this circuit; that is, the output V_{out} of the circuit in Fig. 8-15 is very well regulated.

Where the load current is to be in the order of several amperes, the circuit in Fig. 8-16a can be used to provide a regulated output voltage. The transistor Q, sometimes called the *pass element*, drops the difference in the unregulated dc source voltage V_{in} and the load voltage V_{out}. This total

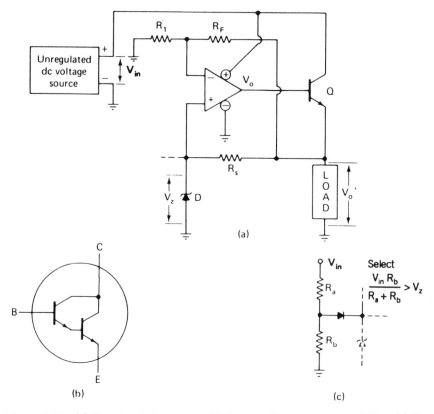

(a)

(b)   (c)

Figure 8-16 (a) Regulated dc source with intermediate current capability; (b) Darlington pair can be used in place of transistor Q for higher load-current capability; (c) voltage divider that can be used to assure starting (cause zener to go into avalanche conduction).

current in the emitter of Q is controlled by a much smaller current in its base. Note that the base current is supplied by the Op Amp.*

This circuit can be analyzed as follows: A current maintained through the zener provides a well-regulated voltage at the noninverting input. This zener's voltage V_z causes an Op Amp output voltage V_o that is A_v times larger. The voltage V_o is applied to the base of the transistor Q and forward-biases its base-emitter junction. For a silicon transistor, this base-emitter drop V_{BE} is typically about 0.7 V. Therefore, the load voltage V_{out} which is the voltage at the emitter, is less than the Op Amp's output V_o by

---

*A 50–200-Ω resistance is sometimes used in series with the base of Q to protect the circuit with excessively low load resistances.

the base-emitter drop V_{BE}; i.e.,

$$V_{out} = V_o - V_{BE}$$

or

$$V_{out} \cong V_o - 0.7 \text{ V}.$$

For most practical purposes,

$$V_{out} \cong V_o.$$

Since the Op Amp is wired to work as a noninverting amplifier,

$$V_{out} \cong A_v V_z.$$

or

$$V_{out} \cong \left(\frac{R_F}{R_1} + 1\right) V_z.$$

In this circuit of Fig. 8.16a, the drop across R_s is the difference in the load and zener voltages, $V_{out} - V_z$. Since V_{out} is regulated, this resistor's drop is quite constant, and therefore, so is the current through it and the zener. A constant current through the zener causes a very constant voltage drop across it. This gives this circuit its good voltage-regulating capability.

REVIEW QUESTIONS 8

8-1. Generally, what is the meaning of the term *linear application* of an Op Amp?

8-2. Since Op Amps are capable of amplifying dc and ac voltages, why should we ever consider using coupling capacitors between Op Amp stages in ac amplifier systems?

8-3. What is the meaning of and the reason for bootstrapping the input of an amplifier?

8-4. If an Op Amp wired as an integrator has a square-wave input voltage applied, what output waveform can we expect?

8-5. If an Op Amp wired to work as a differentiator has a sawtooth input voltage applied, what output voltage waveform can we expect?

8-6. What advantage does the Op Amp give to the circuit of Fig. 8-15 compared to the circuit of Fig. 8-14?

8-7. Why would a circuit like that of Fig. 8-16 be used instead of the circuit of Fig. 8-15?

8-8. In the circuit of Fig. 8-17 what is the purpose of the 240-kΩ resistor?

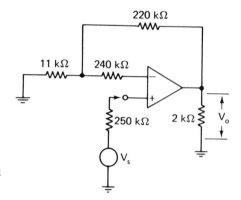

Figure 8-17 Circuit for Problems 8-1 and 8-2.

PROBLEMS 8

Review

8-1. What is the gain V_o/V_s of the circuit in Fig. 8-17?

8-2. In the circuit of Fig. 8-17, if the 11-kΩ resistor is replaced with an 82-kΩ resistor, (a) what is the gain V_o/V_s of the circuit, and (b) with what value should the 240-kΩ resistor be replaced if the circuit stability is to remain close to what it was?

8-3. In the circuit of Fig. 8-18, (a) what is the gain V_o/V_s of the circuit, and (b) what value of resistance R should we use to minimize instability and output offset voltage?

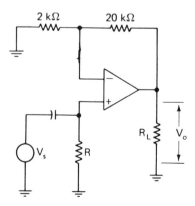

Figure 8-18 Circuit for Problems 8-3 and 8-4.

8-4. If the Op Amp in the circuit of Fig. 8-18 is a 741 type and $R = 1$ MΩ, (a) roughly what maximum output offset voltage, caused by bias currents, could we expect at room temperature? (b) What value of resistor would you use between the inverting input and the connection between the 2-kΩ and 20-kΩ resistors to reduce drift in the output offset?

Section 8.1

8-5. In the circuit of Fig. 8-19, what total gain V_o/V_s can we expect at frequencies at which the reactances of all capacitors are negligible if the potentiometer is adjusted to 0 Ω?

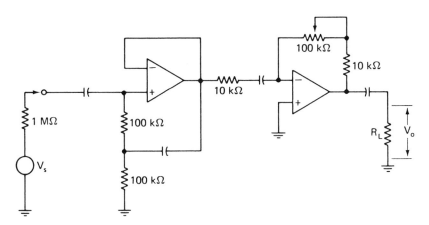

Figure 8-19   Circuit for Problems 8-5 and 8-6.

8-6. At frequencies that cause the reactances of the capacitors to be negligible in the circuit of Fig. 8-19, what gain V_o/V_s can we expect if the potentiometer is adjusted to maximum resistance?

Section 8.4

8-7. In the circuit of Fig. 8-14, V_{in} varies from 15 to 20 V, $I_Q = 3$ mA, $R_1 = 2$ kΩ, and R_2 is a 1-kΩ POT. What is V_{out} when (a) the POT resistance is 800 Ω, and when (b) the POT resistance is 100 Ω?

8-8. In the circuit of Fig. 8-14, V_{in} varies from 20 to 24 V, $I_Q = 4$ mA, $R_1 = 3.3$ kΩ, and R_2 is a 2-kΩ POT. What is V_{out} when (a) the POT resistance is 2 kΩ, and when (b) the POT resistance is 0 Ω?

8-9. Referring to the circuit described in Problem 8-7, if a load current change causes I_Q to increase by 0.5 mA, what is the resulting change in the output V_{out} when the POT resistance is 2 kΩ?

8-10. Referring to the circuit described in Problem 8-8, if a load current change causes I_Q to increase by 0.5 mA, what is the resulting change in the output V_{out} when the POT resistance is 2 kΩ?

Section 8.7

8-11. If, in the circuit of Fig. 8-15, a 3.3-kΩ resistor is placed in series above the 10-kΩ POT while a 1.2-kΩ resistor is placed in series below the 10-kΩ POT, over what range can V_{out} be adjusted, theoretically? *Hint*: The voltage across pin 2 of the regulator and pin 2 of the Op Amp is a regulated 5 V. Also, pins 2 and 6 of the Op Amp are at the same potential.

8-12. If, in the circuit of Fig. 8-15, a 10-kΩ resistor is placed in series above the 10-kΩ POT and a 2-kΩ resistor is placed in series below the 10-kΩ POT, over what range can V_{out} be adjusted, theoretically?

8-13. Referring to the circuit of Fig. 8-16, V_{in} varies from 9 to 12 V and V_{out} is to be regulated at about 5 V. The load draws up to 3 A and $V_z = 3.3$ V. Use an $I_z = 75$ mA. Select the proper ratios of resistors R_F/R_1, of R_a/R_b and the value of R_s. With your component values, what maximum power will the transistor and the zener be required to dissipate?

Section 8.2

8-14. In the circuit of Fig. 8-20, if $R_1 = R_2 = R_3 = 1$ kΩ and if the Op Amp was initially nulled, (a) what is its output voltage V_o, and (b) what value of R_4 should we use to minimize drift?

8-15. If $R_1 = 1$ kΩ, $R_2 = 2$ kΩ, and $R_3 = 4$ kΩ in the circuit of Fig. 8-20, (a) what is its output voltage V_o, and (b) what value of R_4 will minimize drift? Assume that the Op Amp was initially nulled.

8-16. The three input resistors R are to be equal in the circuit of Fig. 8-21. (a) In order to minimize drift, what value should each of these input resistors be? (b) What is the voltage at the noninverting input, and (c) what is the voltage V_o? Assume that the Op Amp was nulled.

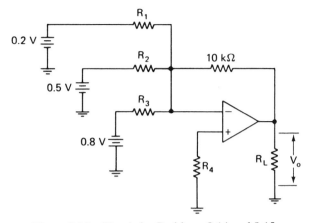

Figure 8-20   Circuit for Problems 8-14 and 8-15.

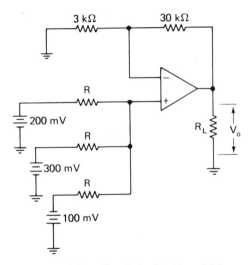

Figure 8-21   Circuit for Problem 8-16.

Section 8.3

8-17. In the circuit of Fig. 8-9, if $R = 100$ kΩ, $C = 100$ μF, and the applied voltage V_s has the waveform shown in Fig. 8-22. What is the output voltage at (a) $t = 1.1$ second and at (b) $t = 5.1$ seconds. Assume that initially the Op Amp was nulled and that the capacitor's voltage was zero.

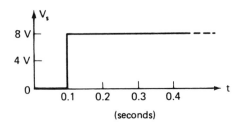

Figure 8-22   Waveforms for the integrator circuit Problem 8-17.

8-18. In the circuit described in Problem 8-17, how long will it take for the amplifier to saturate after the 8-V step voltage is applied, if the output saturation voltages are +16 V and −16 V?

9

ACTIVE FILTERS

Electrical filters were in use long before IC Op Amps were available. Typically they were constructed with passive components: resistors, capacitors, and inductors, and were in fact called passive filters. With low-cost and reliable amplifying (active) devices available, such as Op Amps, filter designs are now mainly active types. For most applications, active filters are easy to design and they eliminate the need for high-cost inductors.

In this chapter we will find that active filters are classified in four categories. These are

1. Low-pass filters
2. High-pass filters
3. Bandpass filters, and
4. Bandstop filters

Low-pass filters allow low frequencies to pass while rejecting higher values. High-pass filters reject low frequencies but pass the highs. Bandpass filters serve to amplify or pass a narrow range of frequencies while attenuating or rejecting all others. And the bandstop filters are used where a narrow range of frequencies is to be rejected while all others are to pass through.

9.1 BASIC RC FILTERS

The series RC circuit of Fig. 9-1 is a simple passive *low-pass* (LP) filter. Its frequency response is shown in part b of this figure. At dc and relatively low

169

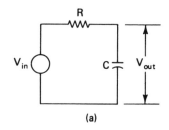

(a)

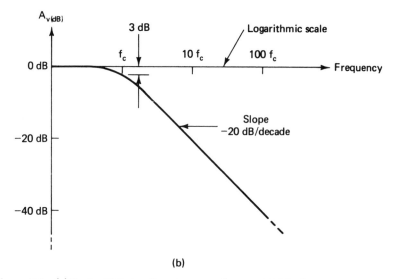

(b)

Figure 9-1   (a) Series RC circuit as low-pass filter and (b) its frequency response.

frequencies (lower than f_c), the output V_{out} is about at the same amplitude as the input V_{in}. Higher frequencies of V_{in}, however, cause the reactance of the capacitor C to decrease. This causes an increased voltage drop across the resistance R and a corresponding decrease in the output voltage V_{out}. Generally,

$$V_{out} = \frac{V_{in}(-jX_C)}{R - jX_C},$$

(9-1a)

where

$$-jX_C = \frac{-j}{2\pi fC} = \frac{-1}{j\omega C},$$

(9-2)

or, neglecting phase shift

$$V_{out} = \frac{V_{in} X_C}{\sqrt{R^2 + X_C^2}} \cdot$$

(9-1b)

Gain A_v is the ratio of the output V_{out} to the input V_{in}. Therefore,

$$A_v = \frac{V_{out}}{V_{in}} = \frac{-jX_C}{R - jX_C}$$

(9-3a)

or

$$|A_v| = \left| \frac{V_{out}}{V_{in}} \right| = \frac{X_C}{\sqrt{R^2 + X_C^2}} \cdot$$

(9-3b)

The phase shift of V_{out} with respect to V_{in} is

$$\phi = -\arctan\left(\frac{R}{X_C}\right) = -\arccos\left(\frac{V_{out}}{V_{in}}\right)$$

In decibels, the gain

$$A_{v(dB)} = 20 \log \left| \frac{V_{out}}{V_{in}} \right| = 20 \log |A_v|.$$

(9-4)

The frequency at which the response curve starts to bend (breaks) is called the *cutoff frequency* f_c, *critical frequency*, or *break frequency*. The gain $A_{v(dB)}$ is 3 dB below its maximum value at f_c (see Fig. 9-1b). Note that at frequencies higher than f_c, the gain $A_{v(dB)}$ rolls off at a 20-dB/decade rate.

The use of resistance and capacitance to filter (remove) higher frequencies is not new to us. We already know that filter capacitors are placed across the

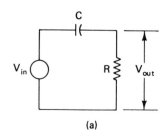

(a)

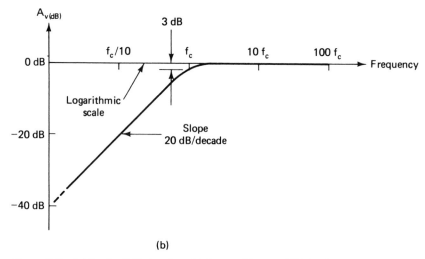

(b)

Figure 9-2 (a) Series RC circuit as high-pass filter and (b) its frequency response.

outputs of dc power supplies to filter out the ripple frequencies and transient pulses. Even the 741 Op Amp has an internal 30-pF capactor that, combined with internal resistance of its circuitry, causes the -20 dB/decade to roll off of its open-loop vs. frequency response curve (examine the circuit and curve of Appendix F).

The RC circuit of Fig. 9-2 works as a passive *high-pass* (HP) filter. At higher frequencies, the reactance of the capacitor C is negligible, causing V_{in} to appear across the resistor R; that is, $V_{out} \cong V_{in}$. At lower frequencies, the increased voltage drop across C causes V_{out} to decrease, as shown in Fig. 9-2b. In this case,

$$A_v = \frac{V_{out}}{V_{in}} = \frac{R}{R - jX_C} \tag{9-5a}$$

or

$$\boxed{|A_v| = \left|\frac{V_{out}}{V_{in}}\right| = \frac{R}{\sqrt{R^2 + X_C^2}} \cdot}$$

(9-5b)

The phase shift of V_{out} with respect to V_{in} is

$$\phi = \arctan\left(\frac{X_C}{R}\right) = \arccos\left(\frac{V_{out}}{V_{in}}\right).$$

As with the LP filter, $A_{v(dB)}$ is 3 dB below its maximum value at the cutoff frequency f_c. As shown, at frequencies lower than f_c, the gain $A_{v(dB)}$ vs. frequency curve has a slope of 20 dB/decade.

In either circuit of Fig. 9-1 or Fig. 9-2, the reactance of the capacitor C is equal to the resistance R at the cutoff frequency f_c. Therefore, since

$$X_C = \frac{1}{2\pi fC},$$

then

$$f = \frac{1}{2\pi X_C C} \cdot$$

And at f_c, where $X_C = R$,

$$\boxed{f_c = \frac{1}{2\pi RC} \cdot}$$

(9-6)

This equation enables us to determine quickly the cutoff frequency of a simple RC filter.

9.2 FIRST-ORDER ACTIVE FILTERS

The simple RC filters of Figs. 9-1 and 9-2 cannot provide consistent filtering if V_{out} must be applied across a low or varying load resistance. An Op Amp can serve to buffer (isolate) the load from the filter, as shown in Fig. 9-3. Both circuits are first-order active filters. Generally, active filters use components or devices that are able to amplify, such as transistors or Op Amps. Passive filters consist of passive components only: resistors, capacitors, and/or

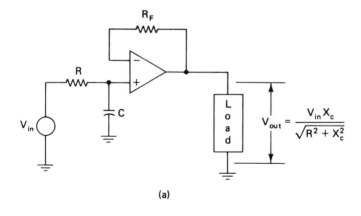

(a)

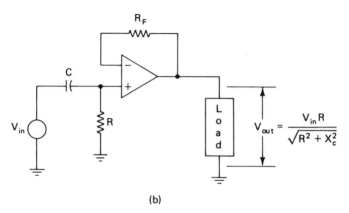

(b)

Figure 9-3 First-order active filters: (a) low pass and (b) high pass (referred to in Review Question 9-15 and in Problems 9-1–9-6).

inductors. A first-order filter has one resistor R and one capacitor C; a second-order filter has two resistors and two capacitors; a third-order filter has three and three, and so on.

Both Op Amps of Fig. 9-3 are wired to work as voltage followers. A resistance $R_F = R$ is used if drifting bias currents cause excessive drift in the quiescent value of V_{out}. Otherwise, R_F is replaced with a short (0 Ω).

Example 9-1

If $R = R_F = 20\ k\Omega$ and $C = 800\ pF$ in the circuit of Fig. 9-3a, find the cutoff frequency f_c. Solve for the gain $A_{v(dB)}$ at (a) $f_c/100$, (b) $f_c/10$, (c) f_c, (d) $10\ f_c$, and (e) $100f_c$.

Answer. By Eq. (9-6),

$$f_c = \frac{1}{2\pi(20 \text{ k}\Omega)800 \text{ pF}} \cong 9947.2 \text{ Hz; call it 10 kHz in round numbers.}$$

Now by Eqs. (9-2), (9-3b), and (9-4), we find that

(a) at $f = 100$ Hz, $A_{v(\text{dB})} \cong -4.4 \times 10^{-4}$ dB $\cong 0$ dB
(b) at $f = 1$ kHz, $A_{v(\text{dB})} \cong -4.4 \times 10^{-2}$ dB $\cong 0$ dB
(c) at $f = 10$ kHz, $A_{v(\text{dB})} \cong -3$ dB
(d) at $f = 100$ kHz, $A_{v(\text{dB})} \cong -20$ dB
(e) at $f = 1$ MHz, $A_{v(\text{dB})} \cong -40$ dB

If you plot these $A_{v(\text{dB})}$ vs. f values on semilog paper, you'll get a curve like that of Fig. 9-1b.

9.3   NTH-ORDER ACTIVE FILTERS

To generalize, we can say that an LP filter passes frequencies below the cutoff frequency f_c and attenuates frequencies above f_c. In most applications, the -20-dB/decade roll-off of $A_{v(\text{dB})}$ is inadequate. Ideally, an LP filter's $A_{v(\text{dB})}$ vs. f curve should roll off (decrease) vertically above f_c (see curve 5 in Fig. 9-4). An ideal curve is difficult to achieve but improvements are quite practical. An improved curve has a roll-off that is steeper than -20 dB/decade.

By cascading two first-order LP filters, we can have a second-order LP filter. Three first-order filters cascaded can yield a third-order filter, etc. As shown in Fig. 9-4, a second-order LP filter has a -40-dB/decade roll-off, and a third-order LP filter has a -60-dB/decade roll-off. Obviously, each additional first-order filter placed in cascade increases the roll-off by another -20 dB/decade. However, since each Op Amp adds noise, the number that can be cascaded is limited. Later we will see how to build a second-order filter with only one Op Amp.

Figure 9-5 shows three first-order LP active filters cascaded. Let's say, for example, that the first stage has a cutoff frequency of f_{c1}, the second stage has a higher cutoff of f_{c2}, and that the third stage has an even higher cutoff of f_{c3}. Their individual $A_{v(\text{dB})}$ vs. f curves are shown in parts a through c of Fig. 9-6. The curve for the circuit as a whole is shown in d of this figure. This

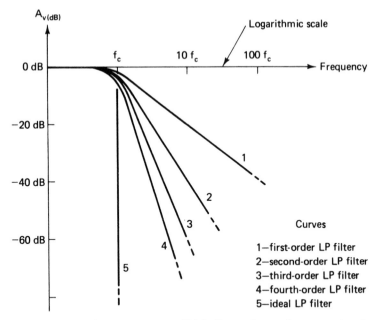

Figure 9-4 $A_{v(dB)}$ vs. frequency curves if LP filters: first order, second order, third order, etc.

overall $A_{v(dB)}$ vs. f curve is the result of superimposing the individual stages' curves. Recall that the uncompensated Op Amp has a similar open-loop vs. frequency curve which is caused by the multiple stages within the Op Amp (see Fig. 6-3).

Note in Fig. 9-6 that at frequencies higher than f_{c3}, the gain of each stage is decreasing at -20 dB/decade, causing an overall $A_{v(dB)}$ roll-off of -60 dB/decade. If all three first-order filters in Fig. 9-5 are designed to

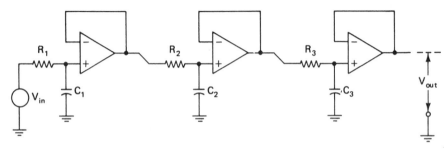

Figure 9-5 Three first-order LP filters cascaded (referred to in Review Question 9-16).

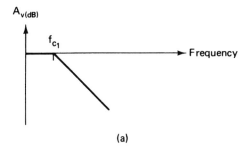

(a)

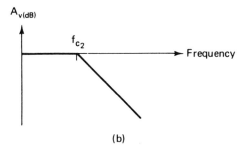

(b)

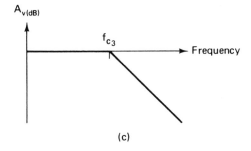

(c)

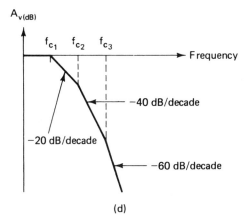

(d)

Figure 9-6 (a)–(c) are $A_{v(dB)}$, vs. frequency curves of three individual LP filters, and (d) their curve when cascaded.

have the same cutoff frequency f_c, the circuit as a whole is a third-order filter. Its $A_{v(dB)}$ vs. f curve rolls off at 60 dB/decade at frequencies above f_c (see curve 3 of Fig. 9-4.)

Example 9-2

Referring to the circuit of Fig. 9-5, if $R_1 = 160$ kΩ, $C_1 = 200$ pF, $R_2 = 270$ kΩ, $C_2 = 50$ pF, $R_3 = 2.7$ kΩ, and $C_3 = 4000$ pF, find the cutoff frequency of each stage. Sketch the overall circuit $A_{v(dB)}$ vs. f curve.

Answer. Using Eq. (9-6), we find that for the first stage, the cutoff frequency

$$f_{c1} = \frac{1}{2\pi(160 \text{ k}\Omega)200 \text{ pF}} \cong 4.97 \text{ kHz}.$$

For the second stage,

$$f_{c2} = \frac{1}{2\pi(270 \text{ k}\Omega)50 \text{ pF}} \cong 11.8 \text{ kHz}.$$

And for the third stage

$$f_{c3} = \frac{1}{2\pi(2.7 \text{ k}\Omega)4000 \text{ pF}} \cong 14.7 \text{ kHz}.$$

The overall $A_{v(dB)}$ vs. f curve should look like the one of Fig. 9-6d. At frequencies lower than f_{c1}, the curve is flat. Approaching higher frequencies, the curve breaks at about 4.97 kHz and rolls off at a 20-dB/decade rate. Approaching still higher frequencies, the curve breaks again and at about 11.8 kHz rolls off at a 40-dB/decade rate. Moving further to the right on the frequency axis, the curve breaks once more at about 14.7 kHz and rolls off at a 60-dB/decade rate.

Example 9-3

Referring again to the circuit of Fig. 9-5, if $R_1 = R_2 = R_3 = 470$ kΩ and $C_1 = C_2 = C_3 = 330$ pF, find the cutoff frequency of each stage and sketch the overall (total) $A_{v(dB)}$ vs. f response curve.

Answer. Each stage has the same cutoff frequency. By Eq. (9-6), we find it is

$$f_c = \frac{1}{2\pi(470 \text{ k}\Omega)330 \text{ pF}} \cong 1026 \text{ Hz.}$$

The curve is flat at frequencies well below f_c. Approaching higher frequencies, it breaks at about 1 kHz and rolls off at a 60-dB/decade rate (see curve 3 of Fig. 9-4).

If first-order *high-pass* (HP) filters are cascaded, as in Fig. 9-7, they can work as a higher-order HP filter. If the resistors and capacitors are selected

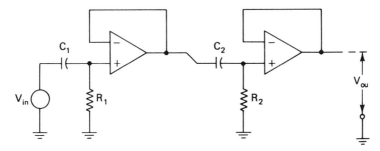

Figure 9-7   First-order high-pass filters cascaded.

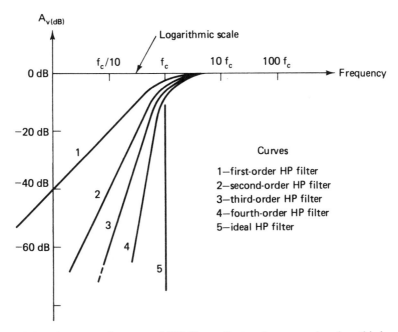

Figure 9-8 $A_{v(\text{dB})}$ vs. f curves of HP filters; first order, second order, third order, etc.

so that both filters attenuate frequencies below the same cutoff frequency f_c, then each stage has an $A_{v(dB)}$ vs. f curve like curve 1 in Fig. 9-8. The overall $A_{v(dB)}$ vs. f response of both stages is illustrated by curve 2. Add a third stage and curve 3 represents the resulting $A_{v(dB)}$ response, etc. Of course, the high frequencies we can expect to pass through an active HP filter are limited by the bandwidth of the Op Amp.

Example 9-4

Referring to the circuit of Fig. 9-7, if $R_1 = R_2 = 100$ kΩ and $C_1 = C_2 = 0.008$ μF, find f_c. Determine the overall gain $A_{v(dB)}$ of this two-stage filter at each of the following frequencies: (a) $f_c/100$, (b) $f_c/10$, (c) f_c, (d) 10 f_c, and (e) 100 f_c.

Answer. By Eq. (9-6),

$$f_c = \frac{1}{2\pi(100 \text{ k}\Omega)0.008 \text{ }\mu\text{F}} = 198.9 \text{ Hz; call it 200 Hz in round numbers.}$$

Now by Eqs. (9-2), (9-5b), and (9-4), we find that

(a) at $f = 2$ Hz, $A_{v(dB)} \cong -40$ dB for each stage. For two stages in cascade, the total $A_{v(dB)} \cong 2(-40 \text{ dB}) = -80$ dB.

(b) At $f = 20$ Hz, $A_{v(dB)} \cong -20$ dB for each stage or a total $A_{v(dB)} \cong -40$ dB.

(c) At $f = 200$ Hz, $A_{v(dB)} \cong -3$ dB for each stage or a total $A_{v(dB)} \cong 2(-3 \text{ dB}) = -6$ dB.

(d) At $f = 2$ kHz, $A_{v(dB)} \cong -0.043$ dB $\cong 0$ dB for each and both stages.

(e) At $f = 20$ kHz, $A_{v(dB)} \cong -0.00043$ dB $\cong 0$ dB for each and both stages.

Plot these on semilog paper and you'll get curve 2 of Fig. 9-8.

9.4   TYPICAL SECOND-ORDER LOW-PASS ACTIVE FILTER

The circuit of Fig. 9-9 shows how a single Op Amp can be used in a second-order LP active filter. With proper selection of components, this circuit has a variety of possible $A_{v(dB)}$ vs. f characteristics (see Fig. 9-10). Note that all curves roll off at -40 dB/decade at frequencies well above the cutoff frequency f_c. As with first-order filters, f_c is determined by the resistor and capacitor values.

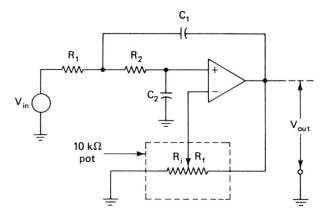

Figure 9-9   Second-order low-pass active filter.

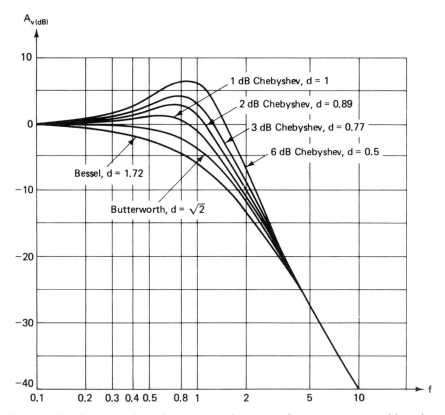

Figure 9-10   Second-order gain $A_{v(dB)}$ vs. frequency f response curves with various damping coefficients d.

The curves of Fig. 9-10 are identified as Bessel, Butterworth, or Chebyshev responses. These are the names of persons whose work contributed to filter technology. Depending on which $A_{v(dB)}$ vs. f curve a given filter has, it is referred to by one of these names.

The Bessel, Butterworth, and Chebyshev filters differ in their damping coefficients d. The Bessel filter has the most damped characteristic, whereas the Chebyshev types are among the least damped. Bessel and Butterworth filters are used where flat, low-frequency (below f_c), and good transient responses are needed. Generally, the Chebyshev filters have better (steeper) initial roll-off but are prone to ringing and overshoot in response to transient or pulsed input signals. The Bessel and Butterworth filters are known for their more linear phase* vs. frequency responses compared to the Chebyshev types.

Each filter has a damping coefficient. As shown in Fig. 9-10, $d = 1.72$ for the Bessel filter, $\sqrt{2}$ for the Butterworth, and so on. The decibel designations on the Chebyshev curves refer to the *ripple channels* of the filters. Figure 9-11 illustrates the ripple channel for 3-dB Chebyshev filters of various orders. In most practical cases, the ripple channels are limited to 6 dB.

The damping coefficient for a second-order LP filter, Fig. 9-9, can be determined with the equation

$$d = \frac{R_1C_2 + R_2C_2 + R_1C_1(1 - A_v)}{\sqrt{R_1R_2C_1C_2}}, \qquad (9\text{-}7)$$

where A_v is the closed-loop gain of the Op Amp.

Its cutoff frequency, in rad/s, is

$$\omega_c = \frac{1}{\sqrt{R_1R_2C_1C_2}}, \qquad (9\text{-}8a)$$

or in Hz

$$f_c = \frac{1}{2\pi\sqrt{R_1R_2C_1C_2}}. \qquad (9\text{-}8b)$$

Obviously, the component values determine the damping coefficient and cutoff frequency, which in turn dictate the shape of the $A_{v(dB)}$ vs. f curve. Change any one value and both d and f_c change also. Filter designers have elaborate tables available that simplify component selection for active filters.

*See Appendix N for equations that determine phase shift caused by second-order active filters.

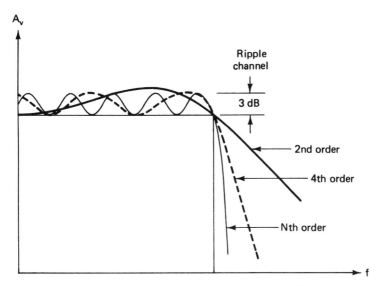

Figure 9-11   Characteristics of 3-dB Chebyshev filters.

For the present, we will simplify matters by limiting the Op Amp's closed-loop gain to unity; that is, $A_v = 1$. In the circuit of Fig. 9-9 then, we can adjust the POT so that $R_f = 0\ \Omega$, or remove the POT and simply wire the Op Amp as a voltage follower. The POT is often used as a trim control. We will also make $R_1 = R_2$. With $A_v = 1$ and $R_1 = R_2 = R$, Eq. (9-8b) becomes

$$f_c = \frac{1}{2\pi R\sqrt{C_1 C_2}},$$

and can be rearranged to

$$R = \frac{1}{2\pi f_c\sqrt{C_1 C_2}}. \tag{9-9a}$$

Similarly, Eq. (9-7) becomes

$$d = 2\sqrt{\frac{C_2}{C_1}},$$

TABLE 9-1. Second-Order Correction Factors

Filter	Correction Factor K
Bessel	0.79 for f_c
Butterworth	1.00 for f_c
1-dB Chebyshev	0.67 for f_p
2-dB Chebyshev	0.8 for f_p
3-dB Chebyshev	0.84 for f_p

and when rearranged, becomes

$$\frac{C_2}{C_1} = \frac{d^2}{4}. \qquad (9\text{-}10)$$

Thus, knowing the type of filter needed, Bessel, Butterworth, or Chebyshev, we determine the damping coefficient from the curves of Fig. 9-10. Equation (9-10) then enables us to select C_1 and C_2 from available capacitor values and types.* After C_1 and C_2 are selected, use Eq. (9-9a) to find the R values for the *Butterworth* filter. For Bessel and Chebyshev types, Eq. (9-9a) is modified to

$$R = \frac{K}{2\pi f_x \sqrt{C_1 C_2}}, \qquad (9\text{-}9b)$$

where K is a correction factor determined from Table 9-1 and $f_x = f_c$ or f_p depending on the type of filter being designed. The correction factor is used because f_c was initially defined as the 3-dB down frequency. However, note in Fig. 9-10 that only the Butterworth filter's curve is at -3 dB at f_c. With the correction factors K listed in Table 9-1, f_x is the 3-dB down frequency for the Bessel filter, whereas f_x is the peak frequency for Chebyshev types. At the peak frequency f_p, the gain $A_{v(\text{dB})}$ reaches its peak. This is just before the $A_{v(\text{dB})}$ vs. f curve enters the -40-db/decade roll-off.

---

*Capacitors with low leakage and good temperature stability should be used; metallized polycarbonate, silvered mica etc.

Example 9-5

We need a second-order LP filter with a $A_{v(dB)}$ vs. f curve that rolls off at -40 dB/decade at frequencies above 4 kHz. Select components for (a) a Butterworth filter and let $f_c = 4$ kHz. Also, select components for (b) a 3-dB Chebyshev filter and let $f_p = 4$ kHz.

Answers

(a) For the Butterworth filter, $d = \sqrt{2} \cong 1.414$. By Eq. (9-10),

$$\frac{C_2}{C_1} = \frac{1.414^2}{4} = 0.5.$$

If we choose $C_1 = 0.02 \ \mu F$, then $C_2 = 0.01 \ \mu F$. By Eq. (9-9a),

$$R = \frac{1}{2\pi f_c \sqrt{C_1 C_2}} = \frac{1}{2\pi (4 \text{ kHz})\sqrt{0.02 \ \mu F(0.01 \ \mu F)}} = 2813.5 \ \Omega.^*$$

(b) For the 3-dB Chebyshev type, $d = 0.77$ and

$$\frac{C_2}{C_1} = \frac{0.77^2}{4} \cong 0.148.$$

If we choose $C_1 = 0.02 \ \mu F$, then $C_2 = 0.148 C_1$, or $0.0029 \ \mu F$. Use $0.003 \ \mu F$. In this case, by Eq. (9-9b)

$$R = \frac{0.84}{2\pi(4 \text{ kHz})\sqrt{0.02 \ \mu F(0.003 \ \mu F)}} = 4.315 \text{ k}\Omega.†$$

9.5 EQUAL-COMPONENT LOW-PASS ACTIVE FILTER

If for a damping coefficient that we need we find it difficult to find available capacitors that satisfy Eq. (9-10), we can take the following approach: Let $R_1 = R_2 = R$ and $C_1 = C_2 = C$ in the circuit of Fig. 9-9. In this case, Eq.

---

*Precision resistors can be used to obtain resistances close to calculated values.
†See Appendix O for BASIC programs that will tabulate $A_{v(dB)}$ vs. frequency characteristics for second-order active filters.

(9-7) simplifies to

$$d = 3 - A_v,$$ (9-11a)

and therefore,

$$A_v = 3 - d.$$ (9-11b)

Also, Eq. (9-8b) becomes

$$f_c = \frac{1}{2\pi RC},$$

and, therefore, for the Butterworth filter,

$$R = \frac{1}{2\pi f_c C}.$$ (9-12a)

For Bessel and Chebyshev filters,

$$R = \frac{K}{2\pi f_x C},$$ (9-12b)

where K is a correction factor obtained from Table 9-1, and as before,

$$f_x = f_c \text{ or } f_p.$$

Equation (9-11b) shows that the Op Amp's voltage gain is *not* unity for the equal-component LP filter. In this case, we use and adjust the potentiometer in the circuit of Fig. 9-12 so that $A_v = 3 - d$ or we can select appropriate fixed resistors R_f and R_i.

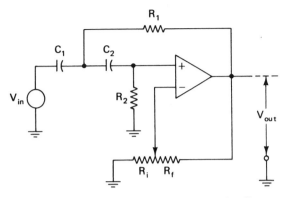

Figure 9-12   Second-order high-pass active filter.

Example 9-6

Design a 3-dB Chebyshev LP filter with a peak frequency $f_p = 4$ kHz. Use the equal-components approach.

Answer. Figure 9-10 shows that $d = 0.77$ for the 3-dB Chebyshev. With Eq. (9-11b) then, we find that

$$A_v = 3 - 0.77 = 2.23.$$

Since the Op Amp in Fig. 9-9 is wired to work as a noninverting amplifier, its gain

$$A_v = \frac{R_f}{R_i} + 1.$$

In this case,

$$2.23 = \frac{R_f}{R_i} + 1$$

or

$$\frac{R_F}{R_i} = 1.23.$$

$R_f = 33$ kΩ and $R_i = 27$ kΩ would do nicely. Next, if we choose $C_1 = C_2 =$

$C = 0.01 \ \mu F$, then, with Eq. (9-12b),

$$R = \frac{0.84}{2\pi(4 \ \text{kHz})(0.01 \ \mu F)} = 3342 \ \Omega; \ \text{use} \ 3.3 \ \text{k}\Omega,$$

where the correction factor 0.84 was read from Table 9-1. If we cannot find standard resistors that are close to calculated values of R, we can experiment with various available capacitor values until R works out more conveniently.

9.6 TYPICAL SECOND-ORDER HIGH-PASS ACTIVE FILTER

The circuit of Fig. 9-12 is a second order HP active filter. With proper selection of its components, its $A_{v(\text{dB})}$ vs. f curve has a 40-dB/decade slope at frequencies below the cutoff frequency f_c. As with LP filters, the component values determine the cutoff frequency and the damping coefficient which determines whether the filter is a Bessel, Butterworth, or Chebyshev type. In this case, the damping coefficient

$$d = \frac{R_1 C_1 + R_1 C_2 + R_2 C_2 (1 - A_v)}{\sqrt{R_1 R_2 C_1 C_2}}, \qquad (9\text{-}13)$$

and as with the LP filter, the HP filter's cutoff frequency

$$f_c = \frac{1}{2\pi\sqrt{R_1 R_2 C_1 C_2}}. \qquad (9\text{-}8b)$$

If we take the equal-component approach—that is, let $R_1 = R_2 = R$ and $C_1 = C_2 = C$—the equations above simplify to

$$d = 3 - A_v, \qquad (9\text{-}11a)$$

which, rearranged, becomes

$$A_v = 3 - d \qquad (9\text{-}11b)$$

and

$$f_c = \frac{1}{2\pi RC},$$

which, for Butterworth filters, can be shown to be

$$R = \frac{1}{2\pi f_c C} \cdot \qquad \text{(9-14a)}$$

If designing Bessel or Chebyshev filters,

$$R = \frac{1}{2\pi f_x CK}, \qquad \text{(9-14b)}$$

where K is the correction factor listed in Table 9-1
   $f_x = f_c$ for the Bessel filter
   or $f_x = f_p$ for the Chebyshev type.

Example 9-7

We need second-order HP filters that significantly attenuate frequencies below 200 Hz. Design (a) a second-order Butterworth filter, letting the cutoff frequency f_c = 200 Hz; and (b) a second-order 1-dB Chebyshev type, letting the peak frequency f_p = 200 Hz.

Answers

(a) Since $d = \sqrt{2} = 1.414$ for the Butterworth filter,

$$A_v = 3 - 1.414 = 1.586;$$

therefore

$$\frac{R_f}{R_i} = A_v - 1 = 0.586.$$

If we select R_f = 3.3 kΩ, then R_i = 5.6 kΩ would suffice. Arbitrarily choosing $C_1 = C_2 = C = 0.2 \ \mu\text{F}$, then

$$R = \frac{1}{2\pi(200 \text{ Hz})0.2 \ \mu\text{F}} \cong 3.978 \text{ k}\Omega. \qquad \text{(9-14a)}$$

(b) Since $d = 1$ for the 1-dB Chebyshev,

$$A_v = 3 - 1 = 2$$

and

$$\frac{R_f}{R_i} = A_v - 1 = 1.$$

If we choose $C_1 = C_2 = C = 0.1\ \mu\text{F}$, then

$$R = \frac{1}{[2\pi(200\ \text{Hz})0.1\ \mu\text{F}]0.67} \cong 11.88\ \text{k}\Omega. \qquad (9\text{-}14\text{b})$$

Of course, the high frequencies that can pass through this filter are limited by the bandwidth of the Op Amp.*

9.7 ACTIVE BANDPASS FILTERS

A *bandpass filter* (BP) permits a range of frequencies to pass through it while it impedes all others. The circuit of Fig. 9-13a is called a multiple-feedback BP filter. With appropriate component values, its $A_{v(\text{dB})}$ vs. f response curve is as shown in part b of this figure. Note that the gain $A_{v(\text{dB})}$ peaks at f_c—the *center frequency*. Also note that $A_{v(\text{dB})}$ is 3 dB below its peak at frequencies f_1 and f_2. The frequencies between f_1 and f_2 are defined as being within the bandwidth (BW); that is,

$$\boxed{\text{BW} = f_2 - f_1.} \qquad (9\text{-}15)$$

The selectivity of a BP filter is usually described by its Q, where

$$\boxed{Q = \frac{f_c}{\text{BW}}} \qquad (9\text{-}16\text{a})$$

or

$$\boxed{\text{BW} = \frac{f_c}{Q}.} \qquad (9\text{-}16\text{b})$$

This last equation shows that larger values of Q yield smaller (narrower) bandwidths. For this circuit of Fig. 9-13, obtainable values of Q lie between 1 and 20. Also, at the center frequency f_c, the gain must be less than twice the

*See Appendix N for equations that determine phase shift caused by second-order active filters.

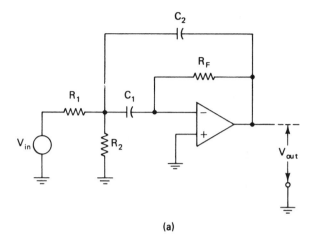

(a)

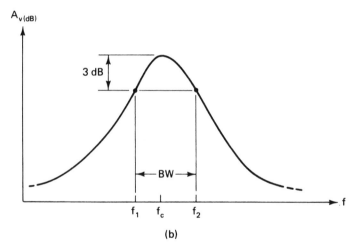

(b)

Figure 9-13 (a) Multiple-feedback bandpass active filter, and (b) its gain vs. frequency response.

Q squared; that is,

$$A_{v_c} < 2Q^2, \qquad (9\text{-}17)$$

where

$$A_{v_c} = \frac{V_{\text{out}}}{V_{\text{in}}} \text{ at } f_c.$$

As with the LP and HP filters, filter design is simplified if we make some of the filter components equal. In this case, if $C_1 = C_2 = C$, then

$$R_1 = \frac{Q}{2\pi f_c A_{v_c} C}, \tag{9-18}$$

$$R_2 = \frac{Q}{2\pi f_c C(2Q^2 - A_{v_c})}, \tag{9-19}$$

and

$$R_F = \frac{2Q}{2\pi f_c C}. \tag{9-20}$$

Example 9-8

Design a BP filter that has a center frequency of 1 kHz and a $Q = 10$. Make the center frequency gain $A_{v_c} = 2$.

Answer. Arbitrarily choosing $C = 0.02\ \mu F$,

$$R_1 = \frac{10}{2\pi(1\ \text{kHz})2(0.02\ \mu F)} = 39.8\ \text{k}\Omega, \tag{9-18}$$

$$R_2 = \frac{10}{2\pi(1\ \text{kHz})0.02\ \mu F\left[2(10)^2 - 2\right]} = 401.9\ \Omega, \tag{9-19}$$

and

$$R_F = \frac{2(10)}{2\pi(1\ \text{kHz})0.02\ \mu F} = 159\ \text{k}\Omega. \tag{9-20}$$

In terms of the multiple-feedback filter's component values, its center frequency

$$f_c = \frac{1}{2\pi}\sqrt{\frac{R_1 + R_2}{R_1 R_2 R_F C^2}}. \tag{9-21}$$

Generally, this circuit's gain A_v can be found with the equation

$$A_v = \frac{(A_v \, d\omega_c \, \omega)^2}{\omega^4 + \omega_c^2(d^2 - 2)\omega^2 + \omega_c^4} \, , \tag{9-22}$$

where

$$\omega = 2\pi f,$$
$$\omega_c = 2\pi f_c,$$
$$d = 1/Q,$$

and

$$A_{v_c} = \frac{R_F}{2R_1} \, , \qquad \text{if } C_1 = C_2. \tag{9-23}$$

The answers to the previous example can be verified with these equations.*

Since the bandwidth is determined by the Q, the 3-dB-down frequencies, which mark the bounds of the BW, can be determined with the following equations:

$$f_1 = f_c\left[\sqrt{\frac{1}{4Q^2} + 1} - \frac{1}{2Q}\right] \tag{9-24a}$$

and

$$f_2 = f_c\left[\sqrt{\frac{1}{4Q^2} + 1} + \frac{1}{2Q}\right]. \tag{9-24b}$$

9.8   ACTIVE BANDSTOP (NOTCH) FILTERS

Often in audio and medical equipment, ac power frequencies are a source of unwanted noise. In such cases, bandstop filters, also called *band-reject* or *notch filters*, can serve to remove the 50-, 60-, or 400-Hz noise while allowing most of the signal frequencies to pass through.

The circuit shown in Fig. 9-14a is a multiple-feedback notch filter. The first stage is a bandpass filter like the one of Fig. 9-13. The second stage is a

---

*See Appendix P for a BASIC program that prints a table of gain vs. frequency characteristics for the multiple-feedback bandpass active filter.

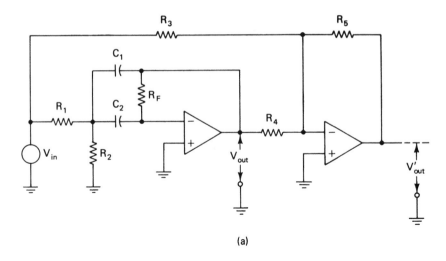

(a)

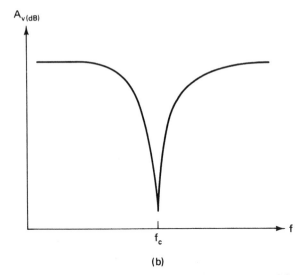

(b)

Figure 9-14 (a) Multiple-feedback bandstop (notch) filter, and (b) its gain vs. frequency response (referred to in Problems 9-21 and 9-22).

summing circuit. Signals V_{in} and V_{out} are applied to the two inputs of this summer. These signals are out of phase and tend to cancel each other's effects via the summing amplifier. Maximum cancellation occurs at the center frequency f_c resulting in a $A_{v(dB)}$ vs. f curve as shown in Fig. 9-14b.

The components R_1, R_2, and R_F in the notch filter are selected as for the bandpass filter; that is, use Eqs. (9-18), (9-19), and (9-20). The ratio R_5/R_4

determines the gain "seen" by signal V_{out} "looking" into the summing amplifier. Similarly, the summer amplifies the signal V_{in} by the ratio R_5/R_3. Maximum cancellation, and greatest depth of the notch, occur at f_c if

$$\boxed{A_{v_c}\left(\frac{R_5}{R_4}\right) = \frac{R_5}{R_3},} \tag{9-25a}$$

where

$$A_{v_c} = \frac{R_F}{2R_1}. \tag{9-23}$$

Equation (9-25a) rearranged becomes

$$\boxed{R_3 = \frac{R_4}{A_{v_c}}.} \tag{9-25b}$$

Resistance R_3 can be a potentiometer for control of the depth of the notch.

Example 9-9

Design a notch filter with a center frequency $f_c = 60$ Hz. Use $Q = 10$ and $A_{v_c} = 2$ in the design Eqs. (9-18)–(9-20).

Answers

Choosing an available capacitor, like $C = 0.1 \ \mu F$, we can show that in the first stage

$$R_1 = \frac{10}{2\pi(60)2(0.1 \ \mu F)} = 132.63 \ k\Omega, \tag{9-18}$$

$$R_2 = \frac{10}{2\pi(60)0.1 \ \mu F\left[2(10^2) - 2\right]} = 1.3397 \ k\Omega, \tag{9-19}$$

and

$$R_F = \frac{2(10)}{2\pi(60)0.1 \ \mu F} = 530.52 \ k\Omega. \tag{9-20}$$

Arbitrarily choosing a ratio $R_5/R_4 = 10$, let $R_5 = 10 \ k\Omega$ and $R_4 = 1 \ k\Omega$.

Then, by Eq. (9-25b),

$$R_3 = 1 \text{ k}\Omega/2 = 500 \ \Omega.^* \qquad (9\text{-}25b)$$

REVIEW QUESTIONS 9

9-1. What is meant by the term *cutoff frequency* as it is used with high-pass and low-pass filters?

9-2. How many resistors and capacitors does a first-order active filter have?

9-3. How do the values of resistance and capacitive reactance compare at the cutoff frequency in a first-order active filter?

9-4. What is the slope of a first-order LP active filter's $A_{v(\text{dB})}$ vs. f curve at frequencies well above the cutoff frequency?

9-5. What is the slope of a first-order HP active filter's $A_{v(\text{dB})}$ vs. f curve at frequencies well below the cutoff frequency?

9-6. What is the slope of a second-order LP filter's $A_{v(\text{dB})}$ vs. f curve at frequencies significantly higher than the cutoff frequency?

9-7. What is the slope of a third-order HP filter's $A_{v(\text{dB})}$ vs. f curve at frequencies significantly below the cutoff frequency?

9-8. How many resistors and capacitors are needed in a second-order LP active filter? How many are needed in a second-order HP active filter?

9-9. A filter section of an audio equalizer consists of two cascaded second-order LP active filters; each has the same cutoff frequency f_c. At what rate does the $A_{v(\text{dB})}$ vs. f curve roll off in the range of $10f_c$?

9-10. What parameter distinguishes the Chebyshev filters from Butterworth types?

9-11. What kind of filter is the circuit of Fig. 9-15b? Specify if it is a first-order, second-order, etc., and if it is a low pass, high pass, bandpass, etc.

---

*See Appendix Q for a BASIC program that prints a table of gain vs. frequency characteristics for the multiple feedback notch filter.

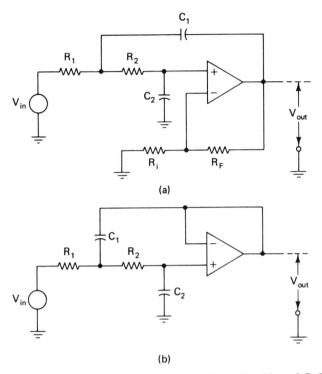

(a)

(b)

Figure 9-15   Review Questions 9-11 and 9-12 and Problems 9-7–9-12.

9-12. Comparing circuits a and b of Fig. 9-15, which is more suited for the equal-component design? Why?

9-13. What kind of filter is the circuit of Fig. 9-16?

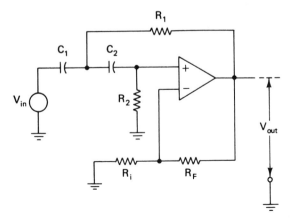

Figure 9-16   Review Question 9-13 and Problems 9-13–9-16.

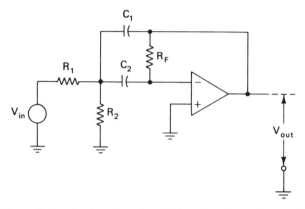

Figure 9-17   Review Question 9-14 and Problems 9-17–9-20.

9-14. What kind of filter is the circuit of Fig. 9-17?

9-15. Sketch modified versions of the circuits in Fig. 9-3 so that each has a gain adjustment.

9-16. How would you modify the circuit of Fig. 9-5 to minimize the drift in the quiescent value of V_{out} caused by drifting bias currents?

9-17. What is meant by the term *bandwidth* in reference to a bandpass filter?

9-18. Define the center frequency in a bandpass filter? In a bandstop filter?

PROBLEMS 9

Sections 9.1 and 9.2

9-1. Referring to the circuit of Fig. 9-3b, if $R = 1.2$ kΩ and $C = 0.22$ μF, what is its cutoff frequency? To minimize drift of V_{out} due to drifting input bias current, what value of R_F should we use?

9-2. Referring to the circuit of Fig. 9-3a, if $R = 1.8$ kΩ and $C = 0.025$ μF in the circuit of Fig. 9-3, what is the cutoff frequency? What value of R_F will minimize the drift of V_{out} due to drifting input bias current?

9-3. Referring to the circuit described in Problem 9-1, what is the gain, in decibels, at frequencies (a) 200 Hz, (b) 400 Hz, (c) 600 Hz, (d) 800 Hz, and (e) 1 kHz?

9-4. What is the gain, in decibels, of the circuit described in Problem 9-2 at each of the following frequencies? (a) 2.5 kHz, (b) 3 kHz, (c) 3.5 kHz, (d) 4 kHz, and (e) 4.5 kHz.

9-5. Sketch an approximate $A_{v(dB)}$ vs. f curve of the circuit described in Problem 9-1. Include the cutoff frequency and assume that frequency is plotted on a logarithmic scale.

9-6. Sketch an approximate $A_{v(dB)}$ vs. f curve of the circuit described in Problem 9-2. Include the cutoff frequency and assume that frequency is plotted on a logarithmic scale.

Sections 9.4 and 9.5

9-7. If, in the circuit of Fig. 9-15a, $R_1 = R_2 = 1.8$ kΩ, $R_i = 1$ kΩ, $R_F = 1.1$ kΩ, and $C_1 = C_2 = 0.047$ μF, what is its damping coefficient? What kind of filter is this?

9-8. If $R_1 = R_2 = 4$ kΩ, $R_i = 5.6$ kΩ, $R_F = 3.3$ kΩ, and $C_1 = C_2 = 0.02$ μF in the circuit of Fig. 9-15a, what is its damping coefficient? What kind of filter is it?

9-9. Referring to the circuit described in Problem 9-7, if you found it to be a Bessel or Butterworth filter, find its cutoff frequency f_c. If you determined that this filter is a Chebyshev type, find its peak frequency f_p. *Hint:* Rearrange Eq. (9-12b) if solving for f_p.

9-10. Referring to the circuit described in Problem 9-8, determine its cutoff frequency f_c, if it is a Bessel or Butterworth type, or find its peak frequency f_p if it is a Chebyshev type. [*Hint:* Rearrange Eq. (9-12b) if solving for f_p.]

9-11. Using the circuit type found in Fig. 9-15b, select components so that it works as a Butterworth LP filter that has a cutoff frequency of 800 Hz. Use $C_1 = 0.2$ μF.

9-12. Using the circuit type a of Fig. 9-15, select components so that it works as a 1-dB Chebyshev LP filter that has a peak frequency of 1 kHz. Use $C_1 = 0.02$ μF.

Section 9.6

9-13. If $R_1 = R_2 = 16$ kΩ, $R_i = 22$ kΩ, $R_F = 27$ kΩ, and $C_1 = C_2 = 0.047$ μF, in the circuit of Fig. 9-16, what is its damping coefficient? What kind of filter is it?

9-14. If in the circuit of Fig. 9-16, $R_1 = R_2 = 20$ kΩ, $R_i = 39$ kΩ, $R_F = 11$ kΩ, and $C_1 = C_2 = 0.1$ μF, what is the damping coefficient? What kind of filter is it?

9-15. Referring to the circuit described in Problem 9-13, determine its cutoff frequency f_c if it is a Bessel or Butterworth type, or find its peak frequency f_p if it is a Chebyshev type. *Hint:* Rearrange Eq. (9-14b) if solving for f_p.

9-16. Referring to the circuit described in Problem 9-14, determine its cutoff frequency f_c if it is a Bessel or Butterworth type, or find its peak frequency f_p if it is a Chebyshev type. *Hint:* Rearrange Eq. (9-14b) if solving for f_p.

Section 9.7

9-17. Select components for the circuit of Fig. 9-17 so that it works as a bandpass filter with a center frequency f_c of 1 kHz, a gain of 10 at f_c and a Q of 10. Use $C_1 = C_2 = C = 0.02$ μF. If your design meets these specifications, what is its bandwidth and at what frequencies does the gain $A_{v(dB)}$ drop 3 dB below its value at f_c?

9-18. Select components for the circuit of Fig. 9-17 so that it works as a bandpass filter with a center frequency f_c of 200 Hz, a gain of 2 at f_c, and a Q of 8. Use $C_1 = C_2 = C = 0.01$ μF. If your design meets these specifications, what is its bandwidth and at what frequencies does its gain $A_{v(dB)}$ drop 3 dB below the gain at f_c?

9-19. Referring to the circuit of Fig. 9-17, if $R_1 = 8.2$ kΩ, $R_2 = 390$ Ω, $R_F = 160$ kΩ, and $C_1 = C_2 = 0.02$ μF, what are its center frequency f_c and gain at the center frequency?

9-20. If, in the circuit of Fig. 9-17, $R_1 = 1.3$ kΩ, $R_2 = 33$ Ω, $R_F = 13$ kΩ, and $C_1 = C_2 = 0.1$ μF, what are its center frequency f_c and gain in decibels at f_c?

Section 9.8

9-21. Select components for the circuit of Fig. 9-14 so that it works as a notch filter with a center frequency of 50 Hz. Use $Q = 5$ and $A_{v_c} = 1$ in the design Eqs. (9-18)–(9-20). Let $C_1 = C_2 = 0.4$ μF and the ratio $R_5/R_4 = 10$.

9-22. Select components for the circuit of Fig. 9-14 so that it works as a notch filter with a center frequency of 400 Hz. Use $Q = 8$ and $A_v = 2$ in the design Eqs. (9-18)–(9-20). Let $C_1 = C_2 = 0.1$ μF and the ratio $R_5/R_4 = 10$.

10

NONLINEAR AND DIGITAL APPLICATIONS OF OP AMPS

In some applications, Op Amps are not required to work in a linear mode. Instead, they are required to repeatedly switch from one output voltage level to another. Occasionally the output voltage swing of an Op Amp is to be kept within very specific limits. In such cases, Op Amps are used with externally wired components that clip output signals that attempt to swing beyond the predetermined limits. There are limitless possible nonlinear applications of Op Amps. The analyses of a few in this chapter will give us insight into others.

10.1 VOLTAGE LIMITERS

The output swing of a general-purpose Op Amp is often too large for the inputs of some circuits. Digital circuits, for example, usually require specific levels of inputs that are not, without modification, available from the output of a typical Op Amp. Several output-voltage-limiting circuits are shown in Figs. 10-1–10-5.

In the circuit of Fig. 10-1, the zener diodes D_1 and D_2 limit the peak-to-peak value of the output voltage V_o. The zener voltage V_{z_1} of D_1 is approximately equal to the maximum possible positive value of V_o—i.e., we cannot drive the output V_o more positive than about the V_{z_1} voltage. The zener voltage V_{z_2} of D_2 similarly is about equal to the maximum negative value of V_o.

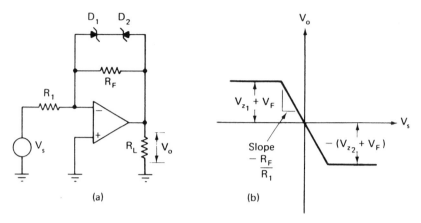

Figure 10-1 (a) Positive and negative voltage clipper and (b) its transfer characteristics.

As shown on the transfer characteristics in Fig. 10-1b, the output V_o is limited to voltage $V_{z_1} + V_F$ on positive alternations. Similarly, note that on negative alternations, V_o is limited to $-(V_{z_2} + V_F)$. The voltage V_F is the voltage drop across the *forward*-biased zener and is typically about 0.7 V. Of course, when V_o swings positively, D_1 is reverse biased and D_2 is forward biased. When V_o swings negatively, D_1 and D_2 are forward and reverse biased, respectively.

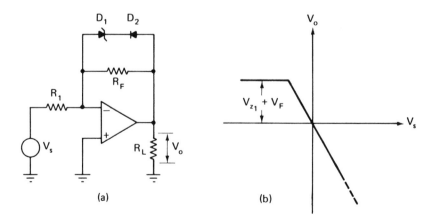

Figure 10-2 (a) Positive voltage clipper and (b) its transfer characteristics.

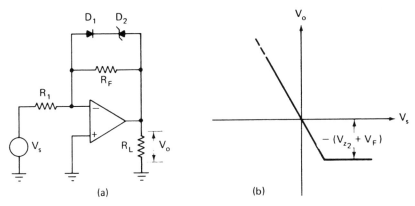

Figure 10-3   (a) Negative voltage clipper and (b) its transfer characteristics.

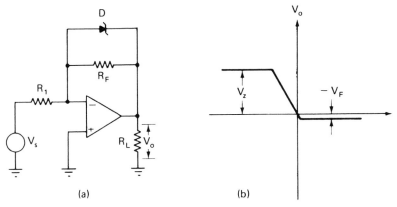

Figure 10-4   (a) Half-wave rectifier with limited positive output and (b) its transfer characteristics.

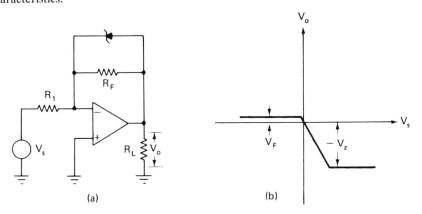

Figure 10-5   (a) Half-wave rectifier with limited negative output and (b) its transfer characteristics.

Example 10-1

In the circuit of Fig. 10-1, suppose that each zener has a 6.3-V zener voltage and a 0.7-V forward drop, and that $R_1 = 1$ kΩ and $R_F = 20$ kΩ. Describe the output waveforms with each of the following sine-wave inputs:

(a) $V_s = 0.3$-V peak,
(b) $V_s = 0.6$-V peak, and
(c) $V_s = 3$-V peak.

Answer. The gain A_v of this stage is about $-R_F/R_1 = -20$, and its output peaks are limited to

$$V_{z_1} + V_F = 6.3 + 0.7 = 7 \text{ V}$$

on positive alternations and to

$$-(V_{z_2} + V_F) = -(6.3 + 0.7) = -7 \text{ V}$$

on the negative alternations.

(a) When $V_s = 0.3$-V peak, $V_o = A_v V_s = -20(0.3) = -6$-V peak. This output V_o does not force either zener into zener (reverse) conduction, and therefore the output waveform is unclipped and sinusoidal.
(b) When $V_s = 0.6$-V peak, the output would peak to $-20(0.6) = -12$ V if the zeners were not across R_F. Since they are, V_o get clipped at 7 V on positive and negative alternations.
(c) When $V_s = 3$-V peak, the output would peak to $-20(3) = -60$ V if limiting factors were not present. Even without the zeners, V_o of a general-purpose Op Amp will run into the rails if we attempt to drive it this hard. With the zeners, the output is clipped at about 7 V on each alternation, and the output V_o has the appearance of a square-wave.

The zener and rectifier diodes in the circuit of Fig. 10-2 limit the swing of the output voltage V_o in a positive direction only, as shown by its transfer

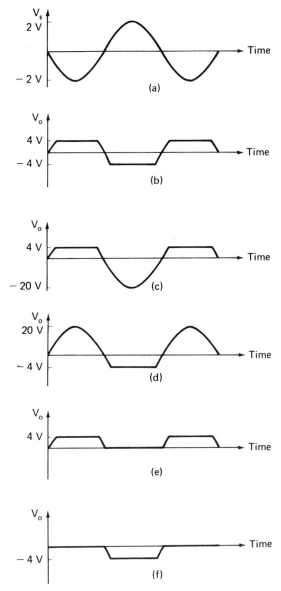

Figure 10-6 If (a) is the input waveform V_s, $R_F/R_1 = 10$, and $V_z = 4$ V, then (b) is the output V_o of the circuit in Fig. 9-1, (c) is the output V_o of the circuit in Fig. 9-2, (d) is the output V_o of the circuit in Fig. 9-3, (e) is the output V_o of the circuit in Fig. 9-4, and (f) is the output V_o of the circuit in Fig. 9-5.

characteristic in Fig. 10-2b. This means that, if V_o attempts to rise above the voltage $V_{z_1} + V_F$, the zener D_1 goes into zener (avalanche) conduction and the waveform of V_o is clipped. The negative alternations are not clipped unless the Op Amp is driven into the negative. The circuit in Fig. 10-3 is similar except that its zener and rectifier diodes limit the output V_o to $-(V_{z_2} + V_F)$ when it swings in the negative direction. V_o can swing positively to the positive rail.

If only a single zener is used with no rectifier, as in Figs. 10-4 and 10-5, and a sine-wave input V_s is applied, the swing in the output V_o is limited by the zener voltage on half of the alternations and by the zener's forward drop V_F on the remaining alternations.

Example 10-2

Referring to the circuits in Figs. 10-1–10-5, suppose that in each, $R_1 = 1\ k\Omega$, $R_F = 10\ k\Omega$, $V_z = 4$ V of each zener, and that the drop across each forward-biased diode and zener is negligible. Sketch the output V_o of each circuit if a 2-V peak, 60-Hz sine-wave input voltage V_s is applied to its input.

Answer. See Fig. 10-6a. This waveform is the input V_s applied to each circuit. Each circuit's gain $A_v = -R_F/R_1 = -10$. Thus the outputs V_o are out of phase with V_s and they *tend* to peak at 20 V on positive and negative alternations. Clipping occurs, however. The waveform shown in Fig. 10-6b is the output V_o of the circuit in Fig. 10-1. The waveform of Fig. 10-6c is the output V_o of the circuit in Fig. 10-2. The waveform in Fig. 10-6d is the output V_o of the circuit in Fig. 10-3. In Fig. 10-6e, the waveform is the output V_o of the circuit in Fig. 10-4. And the waveform in Fig. 10-6f is the output V_o of the circuit in Fig. 10-5.

10.2   THE ZERO-CROSSING DETECTOR

When used in open loop, the Op Amp is very sensitive to changes in input voltages. In fact, fractions of millivolts can easily drive the Op Amp into saturation when no feedback is used. This feature is an advantage in some applications. For example, in Fig. 10-7a, the circuit works as a zero-crossing detector or a sine-wave to square-wave converter. As shown in its output V_o vs. input V_s waveforms in Fig. 10-7b, the output V_o swings into the negative rail when the input V_s passes through zero in the positive direction. On the other hand, when V_s passes through zero negatively, the Op Amp's output V_o is driven into the positive rail.

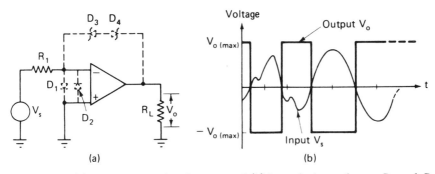

Figure 10-7   (a) The zero-crossing detector and (b) its typical waveforms. D_1 and D_2 are input-protecting diodes and D_3 and D_4 are output-limiting zeners.

If the peak of the input voltage V_s exceeds a volt or so, the input diodes D_1 and D_2 protect the Op Amp. The resistance R_1 limits the current through the input protection diodes. If the Op Amp is a type that is input protected, which means that the equivalent of diodes D_1 and D_2 are built internally, the input diodes are usually unnecessary. If the output swing between $V_{o(max)}$ and $-V_{o(max)}$ is excessive, as it would be for many digital circuit inputs, either or both the zeners D_3 or D_4 can be used to clip V_o at whatever limits are necessary.

10.3   OP AMPS AS COMPARATORS

The comparator, as its name implies, compares two voltages. One is usually a fixed reference voltage V_R and the other a time-varying signal voltage which is often called an *analog* voltage V_A. The Op Amp comparator circuit is very similar to the zero-crossing detector discussed in the previous section, except that a *reference voltage V_R* is used between one input and ground on the comparator (see Fig. 10-8). This causes the Op Amp's output to swing from $V_{o(max)}$ to $-V_{o(max)}$, or vice versa, as the analog voltage V_A passes through the reference voltage value V_R. This reference V_R can be either positive or negative with respect to ground, and of course, its value and polarity determine the V_A voltage that causes the output to switch. Note in Fig. 10-8b that, with the analog input waveform, the output has the waveform in Fig. 10-8c or d, depending on whether V_R is positive or negative, respectively. Obviously, the amplitude of V_A must be large enough to pass through V_R if the switching action is to take place. The circuit in Fig. 10-8 is an inverting type. A noninverting Op Amp comparator and its typical waveforms are shown in Fig. 10-9. A resistor $R_2 = 200 \ \Omega$ or so and one or both zeners D_3 or D_4 are used if necessary to keep the output swing within required limits.

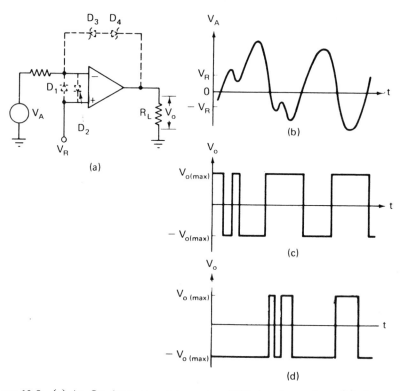

Figure 10-8 (a) An Op Amp as a comparator. With input waveform (b), the output can be waveform (c) with a positive reference voltage, or waveform (d) if the reference is negative.

An Op Amp comparator that has *hysteresis* in its transfer function is shown in Fig. 10-10. In b of this figure, the transfer function shows that the output voltage swings from $V_{o(max)}$ to $-V_{o(max)}$ when the input voltage V_A crosses through V_a while increasing positively. However, note that the output swings from $-V_{o(max)}$ to $V_{o(max)}$ when the input V_A crosses through V_a' when increasing negatively. This comparator, therefore, has two reference voltages; one positive (V_a) and one negative (V_a'), whereas the circuit of Fig. 10-8 has only one reference voltage V_R. The switching of the output V_o at different values of V_A depending on whether V_A is changing in the positive or the negative direction, is called hysteresis and the circuit itself is often called a *Schmitt trigger*.

When the output is at $V_{o(max)}$, the voltage at the noninverting input 2 is positive, having a value

$$V_a = \frac{V_{o(max)} R_a}{R_a + R_b}.$$

(10-1a)

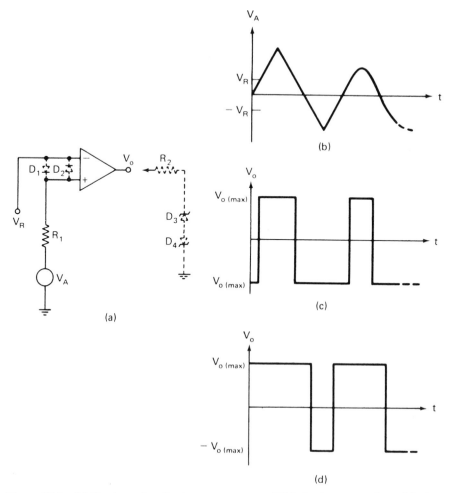

Figure 10-9 (a) Noninverting Op Amp comparator. With the input waveform (b), the output can be waveform (c) if the reference voltage is positive or waveform (d) if the reference is negative.

On the other hand, when the output is at $-V_{o(max)}$, the voltage at input 2 is negative and has the value

$$V_a' = \frac{-V_{o(max)}R_a}{R_a + R_b}.$$ (10-1b)

The difference in the voltages V_a and V_a' is called the *noise margin*. When noise is expected as part of the analog input voltage V_A, hysteresis can effectively eliminate undesirable switching of the output V_o by the noise.

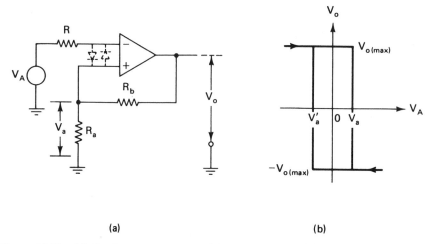

(a)   (b)

Figure 10-10 (a) Op Amp comparator with hysteresis and (b) its transfer function.

Example 10-3

(a) If in the Schmitt trigger circuit of Fig. 10-10, $R_a = 1\ k\Omega$, $R_b = 9\ k\Omega$, the rails are ± 10 V, and the input signal V_A has the waveform of Fig. 10-11, sketch the resulting output.

(b) If this input signal V_A is applied to the circuit of Fig. 10-8 while its reference $V_A = 1$ V, sketch the resulting output.

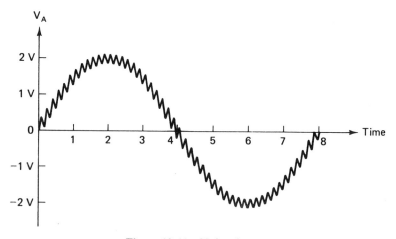

Figure 10-11 Noisy sine wave.

Answers

(a) With the component values specified,

$$V_a = \frac{10\text{ V}(1\text{ kU})}{1\text{ k}\Omega + 9\text{ k}\Omega} = 1\text{ V} \tag{10-1a}$$

and

$$V'_d = \frac{-10\text{ V}(1\text{ k}\Omega)}{1\text{ k}\Omega + 9\text{ k}\Omega} = -1\text{ V}.$$

As shown in Fig. 10-12a, V_o switches to -10 V when V_A crosses through 1 V when going in the positive direction. This output V_o then switches to 10 V when V_A crosses -1 V while changing in the negative direction. The noise margin is 2 V.

(b) Note in Fig. 10-12b that V_o switches *each* time V_A crosses through 1 V and that the noise affects the waveform of V_o.

The threshold voltages at which the Schmitt trigger circuit switches can be shifted by use of a reference voltage V_{ref}, instead of ground, at one end of R_2 (see Fig. 10-13). As indicated by its transfer characteristic shown in Fig. 10-13b, the output is at the positive rail as long as the input signal voltage V_s is less than the *upper threshold voltage* V_a. If V_A rises slightly above this threshold voltage V_a, the output swings to the negative rail and stays there until V_A drops below a *lower threshold voltage* V'_a. The threshold voltages are determined by the components R_1, R_2, and the dc reference voltage V_{ref}. These threshold voltages can be found with the equation

$$\boxed{V_a = \frac{R_2(V_a - V_{ref})}{R_1 + R_2} + V_{ref},} \tag{10-2}$$

where V_o is the maximum positive output voltage when solving for the upper threshold, or

V_o is the maximum negative output voltage when solving for the lower threshold.

Example 10-4

If in the circuit of Fig. 10-13, $R_1 = 10\text{ K}\Omega$, $R_2 = 220\ \Omega$, $V_{ref} = 2$ V, and the rails are ± 10 V, what are the upper and lower threshold voltages?

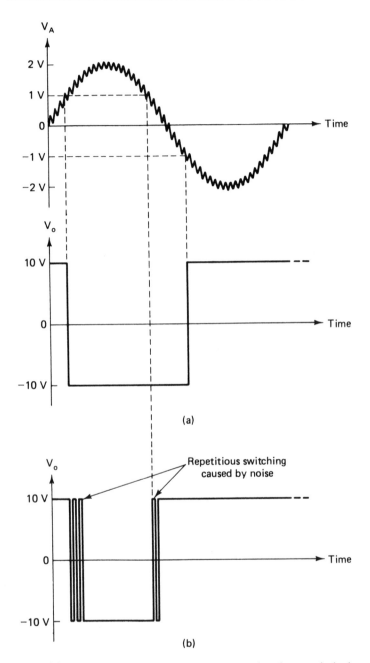

Figure 10-12 (a) Output waveform when comparator has hysteresis in its transfer function, (b) output waveform has repetitious switching caused by noise if comparator has no hysteresis.

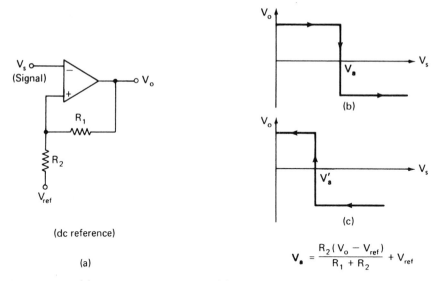

Figure 10-13 (a) Schmitt trigger circuit, (b) transfer function for increasing V_s, and (c) transfer function for decreasing V_s.

Answer. Since the positive rail is $+10$ V, the upper threshold voltage is

$$V_a = \frac{220\ \Omega(10\ \text{V} - 2\ \text{V})}{10.22\ \text{k}\Omega} + 2\ \text{V} \cong 2.17\ \text{V}.$$

This means that if the input voltage V_s rises slightly higher than 2.17 V, the Op Amp is driven into the negative rail, -10 V in this case.

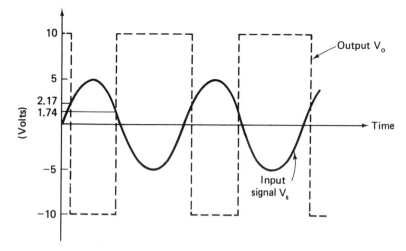

Figure 10-14 Output V_o vs. input V_s waveforms of the Schmitt trigger if the upper threshold voltage $V_a = 2.17$ V and the lower threshold $V_a' = 1.74$ V.

With the output V_o as negative as -10 V, the lower threshold voltage is

$$V_a' = \frac{220 \ \Omega(-10 \ \text{V} - 2 \ \text{V})}{10.22 \ \text{k}\Omega} + 2 \ \text{V} \cong 1.74 \ \text{V}.$$

This means that if the input V_s drops slightly below 1.74 V, the output swings back to $+10$ V. Output vs. input waveforms of this circuit are shown in Fig. 10-14.

10.4   DIGITAL-TO-ANALOG (D/A) CONVERTERS

Digital systems work with two levels of voltage referred to as HIGH and LOW signals or as logic 1 and logic 0. This two-level approach to performing computations and decisions is called a *binary system*, and values in such a system are expressed and processed in binary numbers. Some binary numbers along with their decimal equivalents are shown in Table 10-1. Such binary numbers are often read off of flip-flops followed by level amplifiers which can be viewed here as being equivalent to switches that are capable of providing *either* an output voltage V or 0 V, as shown in Fig. 10-15.

When it is necessary to convert a binary output from a digital system to some equivalent analog voltage, a digital-to-analog (D/A) converter is used.

TABLE 10-1

Decimal	Binary
	DCBA
0	0000
1	0001
2	0010
3	0011
4	0100
5	0101
6	0110
7	0111
8	1000
9	1001
10	1010
11	1011
12	1100
13	1101
14	1110
15	1111
16	10000

Binary output

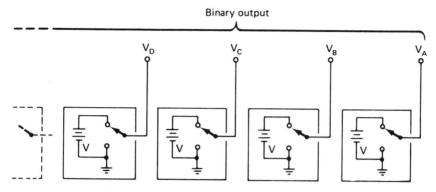

Figure 10-15   Equivalent of a digital system.

An analog output should have 16 possible values if there are 16 combinations of digital voltages V_A through V_D. For example, since the binary number 0110 (decimal 6) is twice the value of the binary number 0011 (decimal 3), an analog equivalent voltage of 0110 is twice the analog voltage representing 0011.

Binary outputs from digital systems can be converted to analog equivalent voltages by the use of either of the resistive networks of Fig. 10-16. The

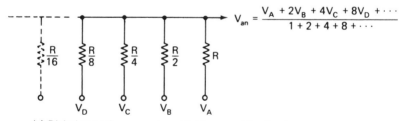

$$V_{an} = \frac{V_A + 2V_B + 4V_C + 8V_D + \cdots}{1 + 2 + 4 + 8 + \cdots}$$

(a) Digital-to-analog converter with binary-weighted resistors.

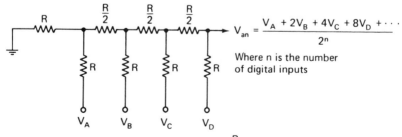

$$V_{an} = \frac{V_A + 2V_B + 4V_C + 8V_D + \cdots}{2^n}$$

Where n is the number of digital inputs

(b) Digital-to-analog converter with R and $\frac{R}{2}$ resistors.

Figure 10-16   Resistive networks as digital-to-analog (D/A) converters.

binary-weighted network in Fig. 10-16a, though requiring fewer resistors, needs a variety of precision resistance values. Its analog output V_{an}, as a function of the two-level inputs, can be determined with the equation

$$V_{an} = \frac{V_A + 2V_B + 4V_C + 8V_D + \cdots}{1 + 2 + 4 + 8 + \cdots}. \qquad (10\text{-}3)$$

If each of its digital inputs, V_A through V_D, is either 15 V or 0 V, the analog outputs V_{an} versus all possible combinations of inputs are shown in Table 10-2.

TABLE 10-2. Output vs. inputs of the binary-weighted resistive network where the digital logic levels are 0 and 15 V.

Decimal equivalent

	V_D	V_C	V_B	V_A	V_{an}
0	0 V	0 V	0 V	0 V	0 V
1	0 V	0 V	0 V	15 V	1 V
2	0 V	0 V	15 V	0 V	2 V
3	0 V	0 V	15 V	15 V	3 V
4	0 V	15 V	0 V	0 V	4 V
5	0 V	15 V	0 V	15 V	5 V
6	0 V	15 V	15 V	0 V	6 V
7	0 V	15 V	15 V	15 V	7 V
8	15 V	0 V	0 V	0 V	8 V
9	15 V	0 V	0 V	15 V	9 V
10	15 V	0 V	15 V	0 V	10 V
11	15 V	0 V	15 V	15 V	11 V
12	15 V	15 V	0 V	0 V	12 V
13	15 V	15 V	0 V	15 V	13 V
14	15 V	15 V	15 V	0 V	14 V
15	15 V	15 V	15 V	15 V	15 V

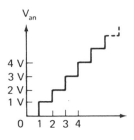

The D/A converter of Fig. 10-16b, using R and $R/2$ resistors, requires more resistors but only two sets of precision resistance values. Its output is

$$V_{an} = \frac{V_A + 2V_B + 4V_C + 8V_D + \cdots}{2^n}, \qquad (10\text{-}4)$$

where n is the n umber of digital inputs. If the levels of each of the inputs are 8 and 0 V, all possible analog outputs of this four-input resistive network are shown in Table 10-3.

If the digital inputs to either network increase consecutively through larger decimal equivalents, the analog outputs are staircase waveforms as shown in Tables 10-2 and 10-3.

TABLE 10-3. Output vs. inputs of the R and $R/2$ resistive network where the digital logic levels are 0 and 8 V.

Decimal equivalent

	V_D	V_C	V_B	V_A	V_{an}
0	0 V	0 V	0 V	0 V	0 V
1	0 V	0 V	0 V	8 V	0.5 V
2	0 V	0 V	8 V	0 V	1 V
3	0 V	0 V	8 V	8 V	1.5 V
4	0 V	8 V	0 V	0 V	2 V
5	0 V	8 V	0 V	8 V	2.5 V
6	0 V	8 V	8 V	0 V	3 V
7	0 V	8 V	8 V	8 V	3.5 V
8	8 V	0 V	0 V	0 V	4 V
9	8 V	0 V	0 V	8 V	4.5 V
10	8 V	0 V	8 V	0 V	5 V
11	8 V	0 V	8 V	8 V	5.5 V
12	8 V	8 V	0 V	0 V	6 V
13	8 V	8 V	0 V	8 V	6.5 V
14	8 V	8 V	8 V	0 V	7 V
15	8 V	8 V	8 V	8 V	7.5 V

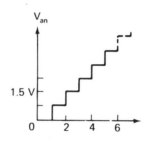

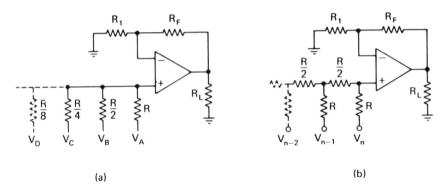

(a)   (b)

Figure 10-17   Digital-to-analog (D/A) converters with buffered output and gain.

Both D/A converters of Fig. 10-16 must work into large-resistance loads; otherwise, the accuracy of their equations, (10-3) and (10-4), degenerates. Therefore, as shown in Fig. 10-17, an Op Amp is well suited to buffer a D/A resistive network to a low-resistance load and to provide gain as well.

Integrated-circuit (IC) D/A converters are available, reliable, and low priced. Op Amps often serve to buffer, or to otherwise condition, the outputs of IC D/A converters. A block diagram of a typical IC D/A is shown in Fig. 10-18. This is a 12-bit *current mode* D/A converter. When used in its full

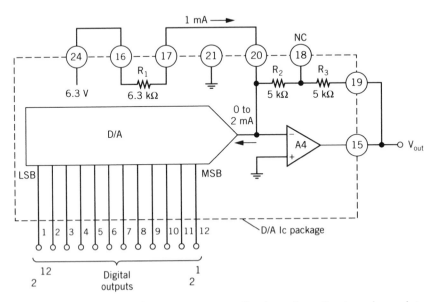

Figure 10-18   Current-to-voltage converter application of an Op Amp in an integrated-circuit current-mode digital-to-analog Converter.

range, its output X will *sink current* from as low as 0 A to as high as 2 mA. In this range there are 4096 (2^{12}) discrete values that correspond with as many possible digital input combinations.

The Op Amp in Fig. 10-18 is built into the same package as is the *current mode* D/A. This Op Amp serves as a *current-to-voltage converter*. In this case, jumpers are placed across its pins 24 and 16 and also across pins 17 and 20. Pin 24 provides a regulated 6.3-V dc output and, in this case, it forces a 1-mA reference current into pin 20 via the internal 6.3-kΩ resistor R_1. The difference in the current that the D/A sinks at output X, and this 1-mA reference current, flows through the series resistors R_2 and R_3. On one extreme, if the 12-bit digital input is 1111 1111 1111, the D/A sinks 0 A. This forces all of the 1-mA reference current through R_2 and R_3; that is, 1 mA flows from left to right through 10 kΩ. The voltage across this 10 kΩ ($R_2 + R_3$), is the output voltage V_o, which is -10 V in this example. On the other extreme, when the digital input is 0000 0000 0000, the D/A sinks 2 mA. Since the reference is only 1 mA, the other 1 mA is pulled from right to left through 10 kΩ and causes a $V_o = +10$ V. A digital value between these extremes is 0111 1111 1111 and it causes the D/A to sink 1 mA. The resulting current through $R_2 + R_3$ is 0 A and the output $V_o = 0$ V.

Example 10-5

Show how to rewire the D/A of Fig. 10-18 so that (a) it will output a V_o that will swing in the range from -5 V to $+5$ V when the digital input varies between 1111 1111 1111, and 0000 0000 0000. And then again rewire the D/A so that (b) it will output a V_o that will swing in the range from 0 V to $+10$ V.

Answers

(a) See Fig. 10-19a. Note that, compared to Fig. 10-18, half of the feedback resistance has been bypassed. Therefore, the range of ± 1-mA current through it will produce an output voltage range of ± 5 V.

(b) See Fig. 10-19b. In this case, the jumper supplying the 1-mA reference is removed. This causes all of the current that the D/A sinks to be pulled, from right to left, through the 5-kΩ of feedback resistance. Since this current ranges from 0 A to 2 mA, the resulting drop across R_2 and output voltage ranges from 0 V to $+10$ V.

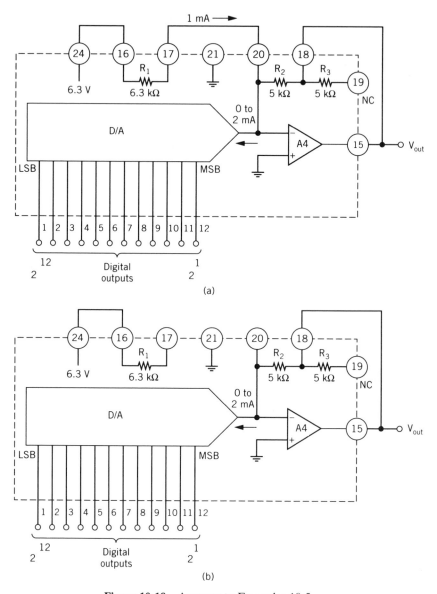

Figure 10-19 Answers to Examples 10-5.

10.5 ANALOG-TO-DIGITAL (A/D) CONVERTERS

With the power and reliability of microcomputers (μC) rising and their prices falling, μCs are increasingly used to monitor *analog events*. We live in an analog world in which physical phenomena change on a continuously variable scale. Temperature changes, increases and decreases in pressures, and variations of volumes of liquids in storage tanks are examples of analog events. In today's competitive world economy, quality of manufactured products is extremely important. Quality is controlled by collecting and using statistics in critical steps of the manufacturing processes. These statistics typically are in analog form and now, increasingly, must be managed by μCs. However, μCs are digital devices and can internally process only digital signals. Therefore, analog signals must be converted to digital format before the μC can process them. The principles of the previously discussed digital-to-analog D/A converter can be reversed to perform analog-to-digital A/D conversions. A system that converts analog voltages into digital equivalent values is shown in Fig. 10-20.

The up-down counter in the system of Fig. 10-20 has a digital output that increases with each clock pulse when its "count-up" line is HIGH and its "count-down" line is LOW. On the other hand, its digital output decreases with each clock pulse when its count-up line is LOW and its count-down line is HIGH.

The Op Amp U_2 works as a comparator. When its output is HIGH, the count-up line is also HIGH. Therefore, when U_2's output is LOW, the count-up line is LOW too. Thus, depending on whether U_2's output is HIGH or LOW, the up-down counter counts digitally up or down, respectively. When the up-down counter is counting up, an upward staircase voltage appears at point a. When the counter is counting downward, a downward staircase is present at point a.

Since the Op Amp U_2 is operated in open loop, its output goes HIGH when its input 1 becomes *slightly* more negative than input 2. Conversely, its output goes LOW when its input 1 becomes *slightly* more positive than input 2.

The first Op Amp U_1 is wired to work as a voltage follower and it buffers the D/A resistive network. Essentially then, the output voltage of the resistive network is applied to input 1 of U_2 and is thus compared with the analog input voltage V_2. This analog input V_2 is the voltage to be digitized.

If the analog input voltage V_2 exceeds the resistive network's voltage V_1, the output of U_2 goes HIGH, and the up-down counter counts up, bringing the network's output voltage V_1 up, in steps, to the analog input V_2. On the other hand, if V_2 is less than or decreases below the voltage V_1, the output of U_2 goes LOW, and the up-down counter counts down, bring voltage V_1 in line with V_2 again.

Generally then, this system has feedback which keeps the voltage output of the resistive network approximately equal to the analog input voltage V_2. Thus, the output of the counter is, or seeks to become, a digital equivalent of the analog input V_2. Though not shown in Fig. 10-20, the counter's outputs can be used to drive a digital readout via an IC latch and decoder. The diodes prevent excessive differential inputs to U_2. The zeners are selected to clip the comparator's output to levels compatible with the up-down counter.

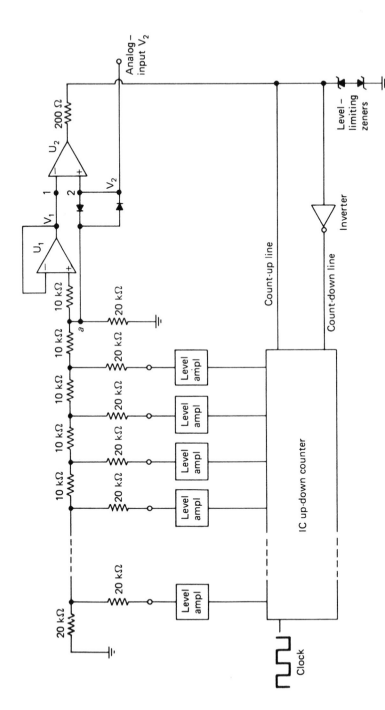

Figure 10-20 Analog-to-digital (A/D) converter.

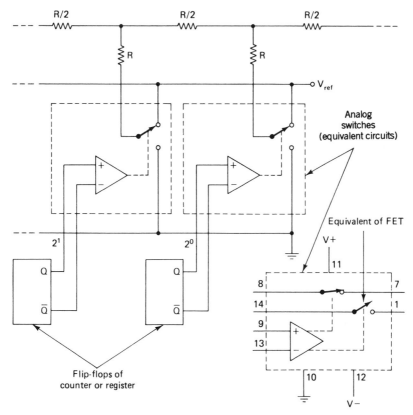

Figure 10-21   Analog switches used to apply either the reference voltage V_{ref} or ground to the inputs of the ladder.

More detailed equivalents of the analog switches are shown in Fig. 10-21. They serve to apply either a regulated reference voltage V_{ref} or ground, to represent logic 1 or 0, to each resistor R. This optimizes the accuracy of the ladder's output.

10.6   ABSOLUTE VALUE CIRCUITS

The output signal V_o of the circuit in Fig. 10-22 can swing positively only, regardless of the polarity of the input signal V_s. Though the input V_s swings through positive and negative values, the output V_o will change proportionally but will vary between zero and positive values. An input V_s and a resulting output V_o for this circuit are shown in Fig. 10-22b and c.

When V_s is positive, diode D_2 is forward biased, and due to the divider action of the equal resistors R_a and R_b, the signal at input 2 is $V_s/2$. Since

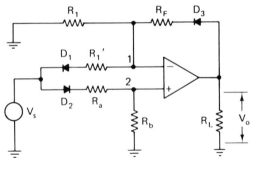

(a) $R_1 = R'_1 = R_F = R_a = R_b$

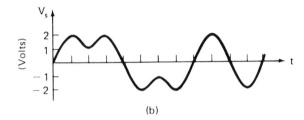

(b)

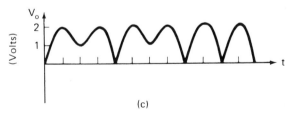

(c)

Figure 10-22 (a) Absolute value output circuit. With input waveform (b) its output has waveform (c).

D_1 is reverse biased, the equal resistors R_F and R_1 form a simple feedback network of a noninverting amplifier, and therefore the output is

$$V_o = A_v \frac{V_s}{2} \cong \left(\frac{R_F}{R_1} + 1 \right) \frac{V_s}{2} \cong V_s.$$

On the other hand, when V_s is negative, D_1 is forward biased and D_2 is reverse biased. Since the left side of R_1 is grounded and now the right side is virtually grounded, it draws negligible current and drops out of consideration

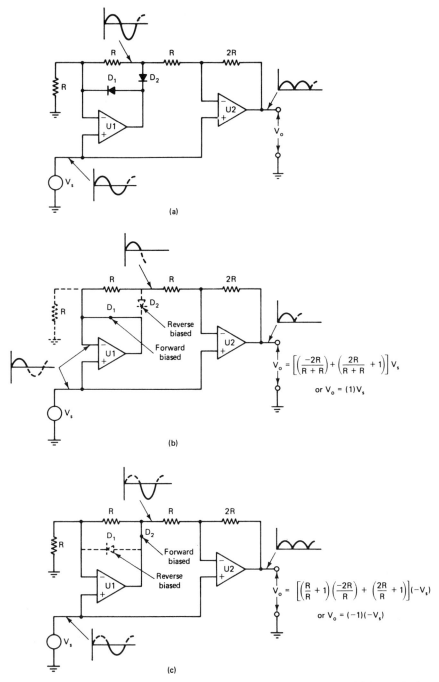

Figure 10-23   (a) Absolute value circuit with high input impedance, (b) its equivalent on positive alternations of V, and (c) its equivalent on negative alternations of V_s.

as far as circuit gain is concerned. Now the output is

$$V_o = A_v V_s \cong -(R_F/R_1') \cong -V_s.$$

Therefore, if resistors R_1, R_1', R_F, R_a, and R_b are all equal, this circuit's gain is either unity or minus unity, depending on the polarity of the input V_s.

The circuit of Fig. 10-23a is an absolute value detector with high input impedance. Its equivalent circuit on positive alternations of the input signal V_s is shown in b of this figure. On negative alternations, the equivalent becomes circuit c. Note the gain equations inside the brackets at the outputs of each equivalent circuit.

In Fig. 10-23a, we see that the input signal V_s has two amplifying paths; one via the noninverting input of U1 and the inverting input of U2. The other path is via the noninverting input of U2 alone. In Fig. 10-23b, the first term of the gain equation represents the gain of V_s via amplifiers U1 and U2. The second term represents the gain of U2 alone. The superposition of these gain paths is the total gain, which is unity (1). Note that on positive alternations of V_s, the far-left resistor R and diode D_2 drop out as far as circuit gain is concerned.

As in Fig. 10-23b, th first term inside the brackets in Fig. 10-23c represents the gain via U1 and U2. The second term is the gain of U2. Overall, as shown, the gain is -1 on negative alternations. In this case, diode D_1 drops out of amplifier action.

10.7   THE OP AMP AS A SAMPLE-AND-HOLD CIRCUIT

The function of the sample-and-hold circuit is somewhat explained by its name. It usually is used to read (sample) an input signal V_s and hold its instantaneous value for a period of time t_H. An Op Amp sample-and-hold circuit is shown in Fig. 10-24. The MOSFET* serves as a switch that is effectively opened or closed by the presence or absence of a control voltage V_c, on its gate G. A positive pulse V_c applied to the gate G of the enhancement-mode MOSFET causes it to become conductive between its drain D and source S leads. This allows the signal V_s to charge the capacitor C. In fact, the voltage across the capacitor follows V_s when V_c is applied. The time periods when pulses V_c are applied are called *sample* periods t_S. The times when gating pulses V_c are not applied are called *hold* periods t_H. During hold periods t_H, the MOSFET is nonconductive, and the charge in the voltage across the capacitor C are held constant. The Op Amp's output is usually read or observed during such times.

The sample-and-hold circuit is typically used to apply to a constant, nonvarying, voltage to the input of an A/D converter. Depending on type,

---

*The MOSFET is a metal-oxide semiconductor field-effect transistor.

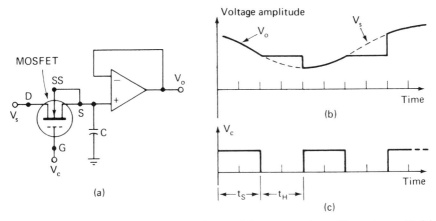

Figure 10-24   (a) A sample-and-hold circuit with output voltage V_o vs. input V_s (b) resulting from the applied control voltage V_c waveform (c).

A/D converters require more or less time to convert an analog input to a digital equivalent. In most A/D designs, the analog input must be constant, while it is converting. Waveforms V_o and V_s in Fig. 10-24b show typical output vs. input waveforms of a sample-and-hold circuit with the control voltage V_C shown in c of this figure.

If the times t_S and t_H are short compared with the time it takes signal V_s to swing significantly, successive readings will yield a close approximation of the output voltage waveform. Of course, since the capacitor's function is to hold a constant voltage for periods of time, it should be a low-leakage type, preferably one with a polycarbonate, polyethylene, teflon or equivalent dielectric.

PROBLEMS 10

Section 10.1

10-1. Sketch the waveform of V_o in the circuit of Fig. 10-1 if V_z of each zener is 6 V, $R_F = 100$ kΩ, $R_1 = 10$ kΩ, and the input signal V_s is sinusoidal with a peak of 1 V. Assume that the Op Amp has negligible offset and that the forward drops of both diodes are negligible.

10-2. In the circuit described in Problem 10-1, sketch the waveform of V_o if D_2 is replaced with a zener whose $V_z = 3$ V.

10-3. Sketch the waveform of V_o in the circuit of Fig. 10-2 if V_z of the zener is 4 V, $R_F = 2120$ kΩ, $R_1 = 5$ kΩ, and the input signal V_s is sinusoidal with a peak of 250 mV. Assume that the Op Amp is nulled and that the forward drops across both diodes are negligible.

10-4. Sketch the waveform of V_o in the circuit described in Problem 10-3 if the 5-kΩ resistor is replaced with 10 kΩ.

10-5. In the circuit described in Problem 10-3, sketch its output V_o waveform if its diode D_2 is replaced with a short.

10-6. Referring again to the circuit described in Problem 10-3, sketch the waveform of V_o if either D_1 or D_2 becomes an open.

10-7. If the zener diode D in the circuit of Fig. 10-4 has a $V_z = 10$ V and the input V_s is sinusoidal with a 5-mV peak, what maximum ratio of R_F/R_1 can we use and still prevent clipping of the output signal's positive alternations?

10-8. Sketch the waveform of V_o that we could expect at the output of the circuit in Fig. 10-5 if its zener's $V_z = 8$ V, $R_F = 220$ kΩ, $R_1 = 1.1$ kΩ, and if V_s is sinusoidal with a peak of 80 mV.

Section 10.3

10-9. The Op Amps in Fig. 10-25 have output limits of $+15$ V and -15 V. With the input voltage V_A shown, what are the output voltages V_{o_1}, V_{o_2}, and V_{o_3} at instants (a) t_2, (b) t_3, and (c) t_5? A_{VOL} of each Op Amp is extremely large.

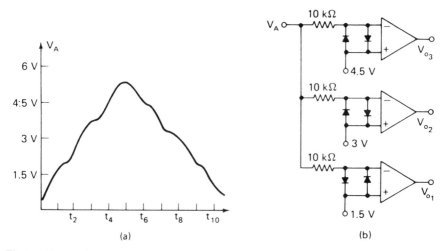

(a)   (b)

Figure 10-25 Comparator circuit (b) with analog input voltage V_A (a) applied for Problems 10-9 and 10-10.

10-10. In the circuit and input described in Problem 10-9, what are the outputs V_{o_1}, V_{o_2}, and V_{o_3} at times (a) t_7, (b) t_9, and (c) t_{10}? Assume that the open-loop gain A_{VOL} of each Op Amp is extremely large.

10-11. Sketch the V_o and V_L waveforms of the circuit in Fig. 10-26b, where the waveform in Fig. 10-26a is the input signal V_s, and $V_R = 0$ V to ground.

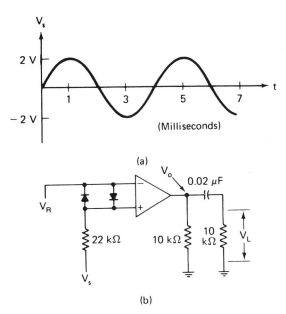

(a)

(b)

Figure 10-26   Circuit for Problems 10-11 and 10-12.

10-12. Sketch the V_o and V_L waveforms of the circuit in Fig. 10-26b, where Fig. 10-26a shows the input signal V_s, and $V_R = +1.4$ V dc to ground.

10-13. If the Op Amp of Fig. 10-10 has output limits of ± 12 V, $R_a = 1$ kΩ, and $R_b = 27$ kΩ, (a) sketch the resulting transfer function. Show the values of V'_a and V_a on the V_A scale. (b) What is the noise margin in volts?

10-14. If the Op Amp of Fig. 10-10 has output limits of ± 12 V, $R_a = 1$ kΩ, and $R_b = 15$ kΩ, (a) sketch the circuit's transfer function. (b) What is the noise margin in volts?

10-15. If the signal of Fig. 10-11 is applied to the input of the circuit described in Problem 10-13, sketch the resulting output waveform.

10-16. If the signal of Fig. 10-11 is applied to the input of the circuit described in Problem 10-14, sketch the resulting output waveform.

Section 10.4

10-17. If the digital input to the circuit of Fig. 10-19a is 0111 1111 1111, how much current will the D/A sink at point X and what will the resulting output voltage be?

10-18. The jumper across pins 17 and 20 is removed and a 6.3-kΩ resistor is placed across these pins. Over what range can the output voltage V_o vary?

Section 10.6

10-19. Suppose that a signal source that has an open-circuit output voltage that is a 3-V peak sine wave and 1 kΩ of internal resistance is used to drive the input of the circuit in Fig. 10-22. If $R_1 = R_1' = R_F = R_b = 1$ kΩ, sketch the resulting output waveform.

10-20. Suppose that a signal source that has 10 kΩ of internal resistance and a 3-V peak sine-wave open-circuit output is used to drive the input of the circuit of Fig. 10-23a. If $R = 2.2$ kΩ, sketch the resulting output waveform.

Section 10.3

10.21. In the circuit of Fig. 10-13, if $R_1 = 40$ kΩ, $R_2 = 10$ kΩ, and $V_{ref} = -1$ V, what are the (a) upper and (b) lower threshold voltages? $V_{o(max)} = \pm 10$ V.

10-22. If $R_1 = 10$ kΩ, $R_2 = 1$ kΩ, and $V_{ref} = +2$ V in the circuit of Fig. 10-13, what are the (a) upper and (b) lower threshold voltages? $V_{o(max)} = \pm 12$ V.

10-23. In the circuit described in Problem 10-21, if the input signal V_s is a 100-Hz sine wave with a peak of 6 V, sketch the output waveform and indicate the approximate width of its positive and negative alternations in milliseconds.

10-24. In the circuit described in Problem 10-22, if V_s is a 400-Hz sine wave with a peak of 10 V, sketch the output waveform and indicate the approximate width of its positive and negative alternations in milliseconds.

Section 10.2 and 10.4

10-25. What is the voltage V_o in the circuit of Fig. 10-27 if the switches 2^0, 2^1, 2^2, and 2^3, are in the 1, 1, 0, and 0 positions, respectively?

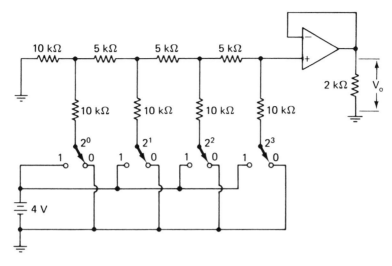

Figure 10-27 Circuit for Problems 10-23 and 10-24.

10-26. What is V_o in the circuit of Fig. 10-27 for each of the following sets of switch positions?

	2^0	2^1	2^2	2^3	V_o
(a)	1	0	0	0	
(b)	1	1	0	0	
(c)	0	0	1	0	

11

OP AMP
SIGNAL GENERATORS

Op Amps can be wired to serve as signal generators capable of a variety of output waveforms. Square waves, triangular waves, sawtooth waves, and sine waves are readily available, to name the more useful waveforms. In this chapter, we will see a few basic Op Amp signal generators and methods of selecting their externally wired components.

11.1 A SQUARE-WAVE GENERATOR

A simple Op Amp square-wave generator is shown in Fig. 11-1. Its output repetitively swings between the positive and negative rails resulting in the square-wave output shown. The time period T of each cycle is determined by the time constant of the components R and C and by the ratio R_a/R_b. This circuit's operation can be analyzed as follows: At the instant the dc supply voltages, $+V$ and $-V$, are applied, zero volts of the initially uncharged capacitor C are applied to the inverting input 1; that is, input 1 is initially grounded. At the same instant, however, a small positive or negative voltage V_b appears across R_b, and this voltage is applied to the noninverting input 2. Voltage V_b initially appears because a positive or negative output voltage V_{oo} exists, even if no differential input voltage is applied to inputs 1 and 2. Thus, the resistors R_a and R_b form a voltage divider, and a fraction of the Op Amp's output voltage is dropped across R_b. Since the inverting input 1 is initially grounded through the uncharged capacitor C, all of the voltage V_b initially appears across the inputs 1 and 2. Even if V_b is small, it will start to drive the Op Amp into one of the rails. If the output offset V_{oo} is positive, the voltage V_b at the noninverting input 2 is positive. This V_b is initially

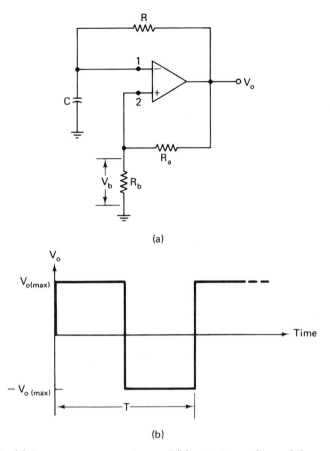

(a)

(b)

Figure 11-1   (a) Square-wave generator, and (b) output waveform of the square-wave generator at low frequencies f where $f = 1/T$.

amplified by the Op Amp's open-loop gain A_{VOL} and drives the output to its limit $V_{o(max)}$, that is, to the positive rail. The rise of $V_{o(max)}$ is at the slew rate of the Op Amp. With the output at $V_{o(max)}$, the capacitor charges through resistor R. If the resistor R and capacitor C formed a simple RC circuit, the capacitor's voltage V_c would eventually rise to $V_{o(max)}$. In this case, however, voltage V_c can rise only to a value slightly more positive than V_b. That is, as V_c rises and becomes a little more positive than V_b, the inverting input 1 becomes more positive than the noninverting input 2, and this drives the output to its negative rail, $-V_{o(max)}$. See Fig. 11-2. With the Op Amp's output at $-V_{o(max)}$, a fraction of this voltage is dropped across R_b. Thus input 2 becomes much more negative than input 1 and holds the Op Amp in negative saturation, at least for a while. The capacitor C now proceeds to discharge and recharge in the negative direction as shown in Fig. 11-2. Now when the

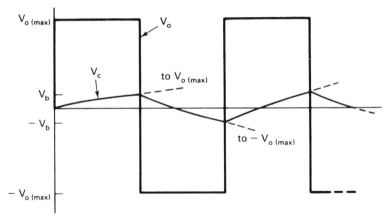

Figure 11-2 Typical output voltage V_o and capacitor voltage V_c waveforms of the square-wave generator.

capacitor's voltage becomes more negative than $-V_b$, the inverting input 1 becomes more negative than input 2, and the output is driven back to $+V_{o(max)}$ to start another cycle.

The *sum* of the resistances R_a and R_b is not critical. It can be selected to be in a broad range, say 10 kΩ to 1 MΩ. A change in their *ratio*, however, does affect the circuit's output frequency f. The period of each cycle is

$$T = 2RC \log_e\left(1 + \frac{2R_b}{R_a}\right)^* \tag{11-1}$$

and the number of cycles per second

$$f = 1/T. \tag{11-2}$$

The smaller the RC product, the faster the capacitor C charges to the voltage across R_b, and the higher the output frequency. Therefore, the resistor R in Fig. 11-1 can be a frequency-selecting potentiometer.

This square-wave generator's output frequency is limited by the slew rate of the Op Amp. If we attempt to operate it at relatively high frequencies, the period of each cycle T is not much larger than the time it takes the output to rise to $V_{o(max)}$ from $-V_{o(max)}$, and vice versa. This causes the generator's output to become trapezoidal or even triangular at higher frequencies.

The peak-to-peak output capability of this square-wave generator can be reduced by means of reduced dc supply voltages or with back-to-back zeners as shown in Fig. 11-3. The zeners will reduce ringing† as well.

---

*Use the ln key on your calculator.
†Ringing refers to an output voltage that oscillates about its eventual steady state after a sudden change.

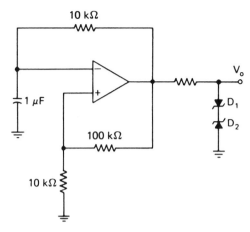

Figure 11-3   Square-wave generator with $\pm V_{o(\max)}$ limited by zeners.

11.2   A TRIANGULAR-WAVE GENERATOR

In Section 8-3 we learned that an integrator's output waveform is triangular if its input is a square wave. See Fig. 8-12a. Apparently, then, an integrator following a square-wave generator, such as in Fig. 11-4, serves as a triangular-wave generator. Since changes in resistance R change the frequency of the square-wave generator's output, the output frequency of the integrator changes. Therefore, if the resistance R is increased or decreased, the frequency of the triangular wave will decrease or increase, respectively. The amplitude of the triangular wave can be controlled somewhat by resistance R_1. Larger or smaller values of R_1 will reduce or increase the output amplitude of the integrator. As with the simple square-wave generator, the output frequency is limited by the Op Amps' slew rates.

11.3   A SAWTOOTH GENERATOR

A sawtooth waveform differs from the triangular waveform in its unequal rise and fall times. The sawtooth can be made to rise positively many times faster than it falls negatively, or vice versa. The circuit in Fig. 11-5 is a sawtooth generator. The first stage is called a *threshold detector*. Its output V_o will swing from the positive to the negative rail when the decreasing output V_o' of the integrator becomes negative enough to pull point x slightly negative with respect to ground. Note that waveforms in Fig. 11-5b. On the other hand, the threshold detector's output swings from the negative to the positive rail when the rising voltage V_o' lifts point x slightly positive with respect to ground. The decrease of V_o' in the negative direction takes longer than its rise in the positive direction because the rate at which the capacitor C charge changes

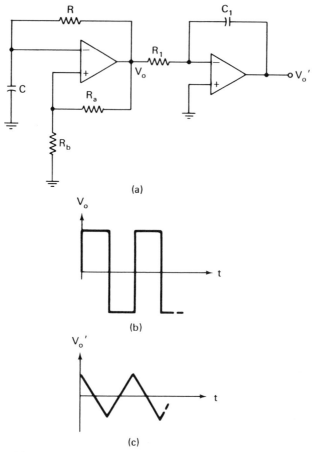

(a)

(b)

(c)

Figure 11-4 (a) Triangular-wave generator, (b) output waveform of the first Op Amp, and (c) output of the second Op Amp.

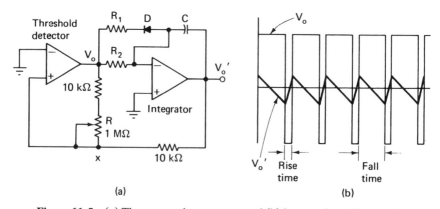

(a)

(b)

Figure 11-5 (a) The sawtooth generator and (b) its waveform if $R_2 > R_1$.

as the polarity of V_o changes. That is, when V_o is at the negative rail, the capacitor C charges mainly through R_1 and the forward-biased diode D. The time constant R_1C is relatively short if R_1 is made significantly smaller than R_2. When V_o is at the positive rail, the capacitor C charges through R_2 more slowly because the time constant R_2C is relatively longer. The values of R_1 and R_2 largely dictate the frequency of the output V_o', while their ratio R_1/R_2 determines the ratio of the rise and fall times. If we reverse the diode D, the rise time of V_o' becomes larger than the fall time.

14.4   VARIABLE-FREQUENCY SIGNAL GENERATORS

The circuit in Fig. 11-6 is an extension of the triangular-wave generator in Fig. 11-4. A range of output frequencies can be selected by the six-position switch shown. Each higher switch position increases the output frequency by 10. Of course, if we intend to use the high-frequency positions 5 and 6, we would select a suitable high-frequency Op Amp with a high slew rate. The

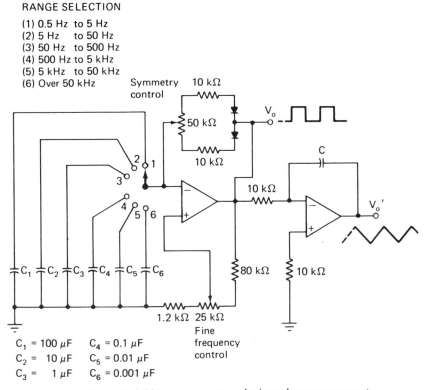

Figure 11-6   Variable square-wave and triangular-wave generator.

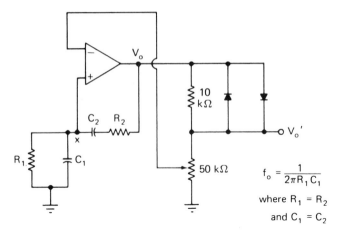

$$f_o = \frac{1}{2\pi R_1 C_1}$$

where $R_1 = R_2$

and $C_1 = C_2$

Figure 11-7   The Wien bridge oscillator.

25-kΩ potentiometer is a fine frequency control. The 50-kΩ symmetry control enables us to vary the ratio of the time of each positive alternation to the time of each negative alternation of the waveforms V_o and V_o'. Thus the output V_o' can be changed from a triangular to a sawtooth waveform by the symmetry control. Capacitor C largely determines the amplitude of V_o'. Generally, the capacitor C must be larger with lower output frequencies. If C is too small, the output V_o becomes clipped because the output Op Amp runs into the rails on each alternation. If C is too large, the amplitude of V_o' is very small, especially at higher frequencies.

11.5   THE WIEN BRIDGE OSCILLATOR

A Wien bridge oscillator is shown in Fig. 11-7. Its output is a sine wave. Note that the Op Amp's output is across a network consisting of R_1, C_1, R_2, and C_2, which is the frequency-selective portion of a Wien bridge. The signal at point x of the Wien bridge network, which is the input signal on the noninverting input, is in phase with the signal V_o at a particular frequency f_c. This frequency is

$$f_c = \frac{1}{2\pi R_1 C_1} \qquad (11\text{-}3)$$

if

$$R_1 = R_2$$

and

$$C_1 = C_2$$

The feedback signal at x leads V_o at frequencies below f_c and lags V_o at frequencies above f_c. Of course then, maximum in-phase feedback occurs at f_c, which therefore is the output frequency of the oscillator. Adjustment of the 50-kΩ potentiometer controls the amount of negative feedback to the inverting input and the amplitude of the output V_o'. The diodes prevent excessive amplitude of negative feedback.

11.6 THE TWIN-T OSCILLATOR

The arrangement of resistors and capacitors, shown in Fig. 11-8a is called a passive twin-T notch filter. With sine-wave voltage input V_{in}, the resistors R_1 and capacitor C_1 (upper T) cause a voltage component at X to lag the input V_{in}. If the frequency at V_{in} is increased, the lagging phase angle between the voltage component at X and V_{in} increases while the amplitude of this voltage component decreases. On the other hand, the capacitors C_2 and resistor R_2 (lower T) cause a voltage component at x to lead V_{in}. If the frequency of V_{in} is increased, this leading phase angle decreases while the amplitude of the voltage component at X increases. The superposition of these two voltage

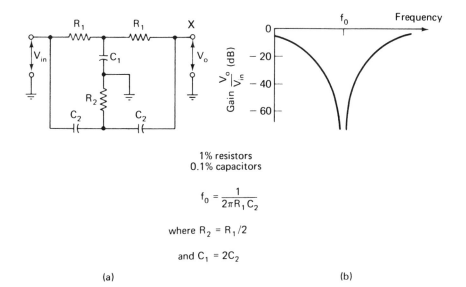

(a)   (b)

Figure 11-8 (a) Passive twin-T notch filter, and (b) typical frequency response of the twin-T notch filter.

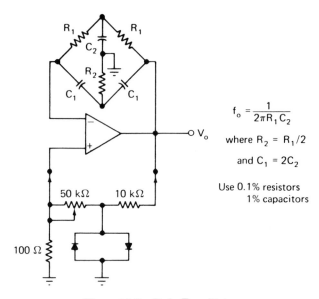

$$f_o = \frac{1}{2\pi R_1 C_2}$$

where $R_2 = R_1/2$

and $C_1 = 2C_2$

Use 0.1% resistors
1% capacitors

Figure 11-9   Twin-T oscillator.

components; one out of the upper T and the other out of the lower T is the output voltage V_o. At what is called the *center frequency* f_c, the two voltage components are out of phase and have equal amplitudes. At f_c, therefore, the two voltage components tend to cancel each other and cause the amplitude of V_o to become minimum. The resulting V_o vs. f curve is as shown in Fig. 11-8b. At the center frequency,

$$\boxed{f_0 = \frac{1}{2\pi R_1 C_2}} \tag{11-4}$$

where

$$R_1 = 2R_2$$

and

$$C_1 = 2C_2.$$

A twin-T sine-wave oscillator is shown in Fig. 11-9. It has two feedback loops. The loop consisting of the 10-kΩ resistor and the 50-kΩ POT causes positive (regenerative) feedback. Positive feedback tends to make the circuit unstable and to go into oscillation. The loop consisting of the twin-T filter is in the negative (degenerative) feedback loop. This filter admits minimum negative feedback and maximum gain at its center frequency f_c causing the circuit to go into oscillation at that frequency.

The component values in the two-T oscillator are critical; 0.1% resistors and 1% capacitors are recommended. The 50-kΩ POT enables us to optimize the waveform and amplitude of V_o.

PROBLEMS 11

Section 11.1

11-1. In the circuit of Fig. 11-1, if $R_a = 100$ kΩ, $R_b = 10$ kΩ, $R = 20$ kΩ, and $C = 1$ μF, what is its approximate output frequency f_o?

11-2. If $R_a = 200$ kΩ, $R_b = 20$ kΩ, and $C = 1$ μF, what value of resistance R should we use in the circuit of Fig. 11-1 if we need approximately a 100-Hz output?

11-3. In the circuit described in Problem 11-1, about what maximum peak-to-peak value of output signal voltage can we expect if the Op Amp is a type 741 and the dc supply voltages are $+15$ V and -15 V?

11-4. If the Op Amp in the circuit of Fig. 11-3 is a type 741 operated with dc supply voltages of $+20$ V and -20 V, and if each zener has a zener voltage $V_z = 6.3$ V, what approximate peak-to-peak output signal voltage can we expect?

Sections 11.2 and 11.3

11-5. If C_1 is too small in the circuit of Fig. 11-4, what effect might it have on the output waveform V_o'?

11-6. If R_1 is too large in the circuit of Fig. 11-4, what effect will it have on the amplitude of the output waveform V_o'?

Section 11.2

11-7. Sketch the approximate waveform of the output V_o' that we could expect from the circuit in Fig. 11-5 if $R_1 = 1$ kΩ and $R_2 = 20$ kΩ.

11-8. Sketch the approximate waveform of the output V_o' that we could expect from the circuit in Fig. 11-5 if the diode becomes open.

Section 11.5

11-9. Select component values for the circuit of Fig. 11-7 so that its output frequency is about 500 Hz. Use $C_1 = 0.2$ μF.

12

POWER OPERATIONAL AMPLIFIERS

The Op Amps and circuits described in the preceding chapters are for low-power applications. Often, transistor power amplifiers are used at the outputs of low-power Op Amps in situations where the signal voltage and/or power levels must be large. Basic power amplifier concepts are described here.

Hybrid IC Op Amps capable of operating with high rail voltages and significant output power values are available. They are called power Op Amps and are used in a variety of applications including audio amplification, motor drives, power supplies, industrial and vehicular controls, telecommunications and test equipment, to name just a few.

A growing segment of the analog electronics industry relies on power Op Amps, their expanding availability, and innovative applications. They are workhorses in the analog electronics world.

A power Op Amp includes, in a single package, what would otherwise require many parts to perform the same function. This reduces design time, printed circuit board (PCB) size, and assembly costs. System reliability and efficiency are additional benefits.

12.1 BOOSTING OUTPUT POWER

The output power capability of an Op Amp can be increased by adding a power amplifier stage at its output, as shown in Fig. 12-1. This output stage consists of a *complementary pair* of power bipolar junction transistors (BJTs). Both of these BJTs are at *cutoff* when the Op Amp's output $V_{out} = 0$ V with

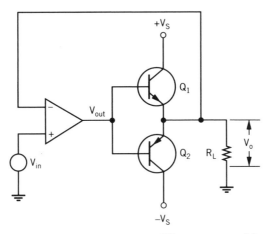

Figure 12-1 Elementary power amplifier at output of Op Amp.

respect to ground. The cutoff resistance of these matched BJTs are equal; therefore, V_o is likewise 0 V when $V_{out} = 0$ V.

When V_{out} swings positively, Q_1 becomes conductive and Q_2 is driven further into cutoff. A conducting Q_1 *sources* current to the load R_L; see Fig. 12-2a. On the other hand, a negative-going V_{out} drives Q_1 further into cutoff and causes Q_2 to conduct and *sink* current as shown in Fig. 12-2b. If V_{out} is ac, Q_1 conducts on positive alternations and Q_2 on negative alternations. Therefore, this output stage is called a *push-pull amplifier*. Transistors that are biased at cutoff, like these, are said to be operating *class B*. (*Class A bias* is about midway between cutoff and saturation.)

12.1-1 Quiescent Power

When $V_o = 0$ V, in the circuit of Fig. 12-1, the power BJTs will draw a relatively small *quiescent current* I_Q from the dc power supplies, and therefore, even with no input or output signal, a power amplifier dissipates power;

$$P_{DQ} = I_Q(+V_S - (-V_S)),\qquad\qquad(12\text{-}1)$$

where P_{DQ} is the quiescent power dissipation,

I_Q is the quiescent current, and

$(+V_S - (-V_S))$ is the total voltage across the power amplifier; this is rail-to-rail voltage and is also called the V_{SS} voltage.

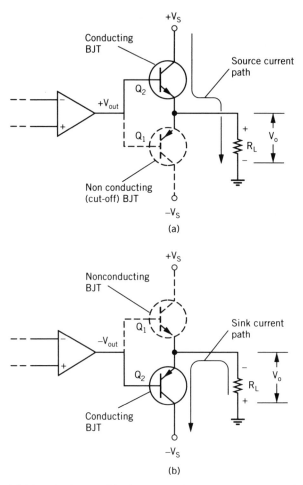

Figure 12-2 (a) V_o swinging positively causes Q_1 to *source current*. (b) V_o swinging negatively causes Q_2 to *sink current*.

12.1-2   ac Power Dissipation

In the circuit of Fig. 12-1, with larger amplitudes of sine-wave inputs V_{in}, the output signal voltage V_o is somewhat sinusoidal too. Actually, this circuit causes *crossover distortion*, which is discussed later. For the present we assume that V_o is a sine wave when V_{in} is sinusoidal and that its peak value is V_p. The total power that the power transistors Q_1 and Q_2 consume is called the amplifier's power dissipation (P_D). The maximum power dissipation ($P_{D(max)}$) occurs when the output voltage has a peak V_p that is about 0.637 times the rail voltage V_S. That is, $P_{D(max)}$ occurs when

$$V_p = 0.637 V_S. \tag{12-2}$$

More precisely, this equation is also shown

$$V_p = \frac{2V_S}{\pi}. \tag{12-3}$$

Since power is voltage squared divided by resistance, this equation leads us to the following for maximum power dissipation:

$$P_{D(max)} = \frac{2(V_S/\pi)^2}{R_L} = \frac{2V_S^2}{\pi^2 R_L}. \tag{12-4}$$

Being able to find a given power amplifier's $P_{D(max)}$ is important. Power amplifier packages are listed by their maximum power dissipating capabilities and heat sinks, too, are selected by this ($P_{D(max)}$) parameter. Although interesting, the derivation of these equations is mathematically lengthy and therefore is shown in Appendix U instead of here. Some power amplifier references show that the maximum power dissipation can also be found with the following equivalent equation:

$$P_{D(max)} = \frac{V_{SS}^2}{2\pi^2 R_L} \cong \frac{V_{SS}^2}{20 R_L}, \tag{12-5}$$

where V_{SS} is the rail-to-rail voltage.

The curve of Fig. 12-3 shows that if the peak of the ac output voltage V_o is either larger or smaller than $2V_S/\pi$, the amplifier's power dissipation is less than maximum.

For reasons discussed later, distinctions are made between ac and dc signal applications of power amplifiers. Generally, an amplifier can safely dissipate more power in ac applications than in dc. Also, signals are consid-

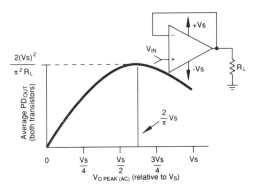

Figure 12-3 Power dissipation vs. ac (> 50 Hz) output voltage V_p. Courtesy of Apex Microtechnology Corporation.

ered to be ac if they are at frequencies of 60 Hz or higher. Because greater thermal stresses occur when operating frequencies are low, some power amplifier manufacturers play it safe and specify that ac is about 200 Hz or higher.

Example 12-1

The circuit of Fig. 12-1 is to amplify ac through a wide range of audio frequencies. The rails are ± 30 V dc, $R_L = 20$ Ω and the quiescent current $I_Q = 5$ mA. Find (a) the maximum average power that both BJTs must be capable of dissipating and (b) the quiescent power dissipation.

Answers

(a) By Eq. (12-4), we find that both BJTs must be capable of dissipating

$$P_{D(max)} = \frac{2(30 \text{ V})^2}{\pi^2 20 \text{ }\Omega} \cong 9.12 \text{ W}.$$

It follows than that each BJT must be able to dissipate at least half of this power, or 4.56 W.

(b) With Eq. (12-1), the total quiescent power

$$P_{DQ} = (5 \text{ mA})(60 \text{ V}) = 0.3 \text{ W}.$$

12.1-3 dc Power Dissipation in the BJTs

In many applications of power amplifiers, a continuous dc current must be delivered to the load R_L. In such cases, the thermal stress on each output transistor is to be greater than in ac applications, where the two push-pull transistors share the load by working alternately. From an engineering point of view, the thermal resistance within the semiconductor of a power amplifier is viewed as a smaller value in ac applications compared to dc.

From electrical fundamentals, we know that the maximum power dissipation in the internal resistance R_i of a dc source occurs when it is equal to its load resistance R_L. Of course, then, when $R_i = R_L$, the voltage across either R_i or R_L is one-half of the open-circuit voltage or, in the case of the split supply circuit of Fig. 12-1, it is either rail voltage $\pm V_S$. In this circuit, as the load R_L "sees" it, the power amplifier is a dc source. Therefore, maximum power is delivered to the load and maximum power dissipation takes place in

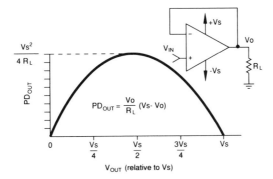

Figure 12-4   Power dissipation vs. dc or low-frequency (< 50 Hz) output voltage V_{out}. Courtesy of Apex Microtechnology Corporation.

the amplifier when

$$V_o = \frac{V_S}{2}.$$

It follows that its maximum power dissipation

$$P_{D(\text{max})} = \frac{(V_S/2)^2}{R_L} \tag{12-5a}$$

or

$$P_{D(\text{max})} = \frac{V_S^2}{4R_L}, \tag{12-5b}$$

where V_S is either rail voltage.

Thus if the voltage V_{out} in Fig. 12-1 is dc or low frequency, less than about 60 Hz, the power amplifier stage has a power dissipation curve as shown in Fig. 12-4. And because dc power amplifiers are often required to output a $V_o = \pm V_S/2$ for prolonged periods of time, each BJT must be capable of dissipating at least this $P_{D(\text{max})}$.

Example 12-2

In the power amplifier stage of Fig. 12-1, signal V_{in} varies in response to temperature changes in an industrial oven. The rails are ± 30 V dc and $R_L = 20 \ \Omega$. Find the maximum power that each BJT must be capable of dissipating.

Answer. By Eq. (12-5b),

$$P_{D(max)} = 30 \text{ V}^2/4(20 \text{ }\Omega) \approx 11.25 \text{ W}.$$

Note that this is significantly larger than required for the same power amplifier in the ac application of Example 12-1.

12.2   SAFE OPERATING AREA (SOA)

Safe operating area (SOA) curves, like the one in Fig. 12-5, show the power handling limitations of a specific amplifier. On linear scales, these current vs. voltage characteristics are curved. They are straight here because both the vertical and horizontal scales are logarithmic. Note that the vertical axis represents current out of either dc supply ($\pm V_S$). This is the collector current I_C of the conducting BJT; see Fig. 12.2. The horizontal axis specifies $V_S - V_o$ voltage. This is the conducting BJT's collector-to-emitter voltage V_{CE}. Note, in Fig. 12-2, that the difference in the dc supply voltage and output voltage is across the conducting BJT. Of course, the product of V_{CE} and I_C is the conducting BJT's power dissipation P_D. This product is constant on the straight portions of any one of these SOA curves. The nonconducting BJT, although having up to the sum of the rail voltages across it, dissipates little power because its leakage current is small. Recall that with ac, these BJTs work in push-pull; that is, when one is conducting, the other is cut off and vice versa. Therefore, the power dissipating capability of the complementary-pair output BJTs must be equal.

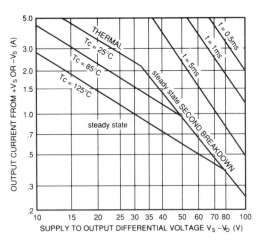

Figure 12-5   Typical safe operating area (SOA) curves. Courtesy of Apex Microtechnology Corporation.

In Fig. 12-5, SOA is to the left of the appropriate curve. Operation on the right is a *SOA violation*. The three far left curves represent steady-state conditions. For example, if $V_{CE} = 30$ V, the BJT can safely conduct a maximum $I_C \approx 2.2$ A at a case temperature $T_C = 25°C$. And that the $I_{C(max)} \approx 1.5$ A at 85°C or 1 A at 125°C. Of course, when dissipating power, the output BJT's junction and case temperatures increase and this lowers (*derates*) $I_{C(max)}$. Generally, as T_C increases, the SOA curve shifts left. The case temperature T_C is very dependent on the type of heat sink used.

Regardless of case temperature, operation on the *thermal* or *secondary breakdown* curves must be avoided. Secondary breakdown is a characteristic of BJTs. When stressed by high V_{CE} and I_C, nonuniform current densities and *hot spots* occur in the BJT junction. These hot spots cause a resulting *thermal runaway*; that is, an avalanche of charge carriers occurs that further increases the temperature. This process is cumulative and leads to failure of the amplifier.

Power amplifiers are often used to output short (time) pulses. The top three curves show that shorter pulses permit operation higher on the I_C vs. V_{CE} characteristics. These pulse curves specify a duty cycle, typically 10%.

Example 12-3

A power amplifier has the characteristics shown in Fig. 12-5. What is its maximum safe power dissipation $P_{D(max)}$ at each of the following case temperatures: (a) $T_C = 25°C$, (b) $T_C = 85°C$, and (c) $T_C = 125°C$.

Answer. Since the power is constant on each of these curves, we can project up from any convenient *safe* value of voltage to each curve and then read the current at the intersection of this projection and the curve. Using, say 30 V, then projecting up from 30 V to

(a) 125°C curve, we intersect it at 1 A. At this temperature, then, $P_{D(max)} = 30$ W.

(b) Similarly, 30 V intersects the 85°C curve at 1.5 A. Thus, $P_{D(max)} = 45$ W.

(c) At the intersection of 30 V and the 25°C curve, we read about 2.2 A. Therefore, $P_{D(max)} \approx 66$ W.

12.3 CURRENT LIMITING ON POWER AMPLIFIERS

Referring back to the simple power amplifier stages shown in Fig. 12-2, the current it sources to or sinks from the load is limited only by the value of this load's resistance R_L. If this load resistance is too low or if the output

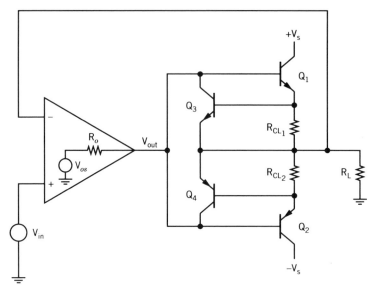

Figure 12-6   Power amplifier with current limiting (CL).

becomes shorted to ground, the current can be excessive and push the BJTs into SOA violation. To keep the power BJTs within their SOAs, current limiting circuitry can be added as shown in Fig. 12-6.

Note that in this circuit of Fig. 12-6, the current limiting resistors R_{CL_1} are in the path of the load current. When Q_1 is sourcing current to R_L, the resulting drop across the current limiting resistor R_{CL_1} places forward bias on the base-emitter junction of Q_3. If this bias voltage is under 0.65 V, Q_3 is nonconducting and the entire value of V_{out} is applied to the base of Q_1. If the load current increases, causing the drop across R_{CL_1} to rise to 0.65 V, Q_3 starts to conduct and thus acts to clamp V_{out}. This prevents Q_1 from sourcing any more current; that is, the load current is limited to a value that causes about a 0.65-V drop across R_{CL_1}.

Similarly, when Q_2 is sinking current, R_{CL_2} applies forward bias on the base-emitter junction of Q_4. If this forward bias rises to -0.65 V, Q_4 starts to conduct and clamps negative-going values of V_{out}. Thus Q_2 is prevented from sinking excessive current.

The value of either current limiting resistor in Fig. 12-5 is selected by the Ohm's law equation

$$R_{CL} = \frac{0.65 \text{ V}}{I_{LIM}}, \qquad (12\text{-}6)$$

where I_{LIM} is the maximum current that the BJT can conduct and still work in its SOA. The power rating of this current limiting resistor must exceed

$$P_{R_{CL}} = 0.65 \text{ V } (I_{LIM}). \tag{12-7}$$

12.4 EXTERNAL CURRENT LIMITING / INTERNAL TRANSIENT PROTECTION

Commercially available power Op Amp packages are available; see Fig. 12-7. The benefits of single package power amplifiers are many. Note in Fig. 12-7 that current limiting resistors R_{CL} are wired externally (outboard), allowing the system designer to select I_{LIM}. Note also the internal diodes between pins 2 and 3 and between 6 and 8. These serve to protect the amplifier from *inductive kickback*. Inductive loads tend to create high-voltage transients when the load voltage is changed suddenly. These transients (*kickback pulses*) are suppressed by the diodes. They provide low-resistance paths to currents that are forced to flow by the collapsing magnetic fields of a load's inductance. Protecting diodes are commonly added to Op Amps that do not have them available internally or to supplement the internal ones; see Fig. 12-8. *Fast recovery* diodes require little time to switch from a conducting

EXTERNAL CONNECTIONS

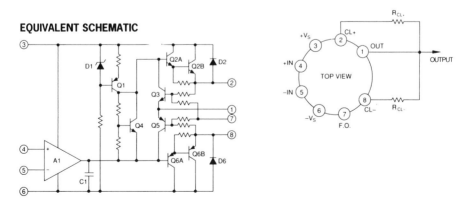

Figure 12-7   Power Op Amp with external pins for current limiting resistors and with internal diodes to protect the output transistors from transients caused by inductive loads.

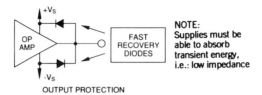

OUTPUT PROTECTION

Figure 12-8 Conventional Op Amp with externally connected diodes to protect the output transistors from transients caused by inductive loads.

state to nonconducting and therefore are recommended for suppressing inductive kickback, especially in high-frequency pulsed applications.

12.5 STABILITY

An amplifier that tends to oscillate is *unstable*. A circuit is oscillating if it has internal parasitic currents. Instability is the most common cause of problems with power Op Amps. Some basic precautions that must be taken to ensure stability are

1. The power supplies must be bypassed (decoupled); see Fig. 12-9. The decoupling capacitors should be close (local) to the amplifier. As shown, small electrolytic (polarized) and ceramic capacitors in parallel are recommended. Use a lot of decoupling on printed circuit boards (PCBs) with multiple amplifiers.

2. Each amplifier must be appropriately grounded. Improper grounding can cause oscillations near the unity-gain-bandwidth (UGBW) frequency of the amplifier, especially with faster Op Amps. A star connection, shown in Fig. 12-9, is recommended. A well-grounded case acts like a Faraday shield, thus minimizing positive feedback by induction.

3. On the PCB, output traces should be not be routed close to the input traces. Close traces tend to be capacitively coupled, resulting in cross talk that can cause oscillating currents. dc power supply traces should be as wide as possible. Resistance in these traces modulates the supply

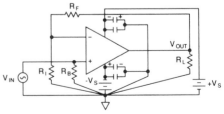

Figure 12-9 Power supply decoupling capacitors and star connected grounding.

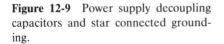

voltage and tends to couple signals between amplifiers. Long traces have inductance that can form high Q resonant (oscillating) circuits when loads are capacitive.

4. In earlier chapters on small-signal Op Amps, we learned to use a resistor R_2 between the noninverting input and ground to minimize the influence of bias current I_B and its drift (see Section 4.2). This resistance is usually not used with power Op Amp designs. It can act like a high-resistance antenna and pick up positive feedback that causes oscillations. Since power Op Amps are hybrids that have relatively small bias currents, removing R_2 has minimal effect.

5. Generally, impedances/resistances are to be kept at a minimum with power Op Amps, especially in high-frequency/high-speed applications. As mentioned in the previous paragraph, high resistances tend to be noise sources.

12.6   INPUT COMPENSATION

Compensating components on the input, shown as R_n and C_n in Fig. 12-10, can prevent oscillations in an amplifier if the cause is in the local feedback loop (reactive load, phase lag at higher frequencies, etc.). The impedance of these components is low at high frequencies. This reduces the feedback amplitude at higher frequencies that would otherwise cause oscillations. Refer back to Section 6.2 for a discussion of phase lag at higher frequencies. If the oscillations are caused by local feedback loop problems, that is, not caused by improper dc supply bypassing, ground loops, etc., the following empirical method can be used to select R_n and C_n:

1. Place a large resistance R_n across the inverting and noninverting inputs. Reduce it until oscillations cease.

2. Place a capacitor C_n in series with R_n. If oscillations resume, increase C_n until they stop. If oscillations do not resume with your introduction of C_n, reduce its value until they do. Then increase C_n to a value that is just large enough to stop the oscillations.

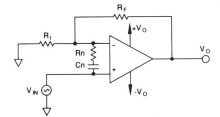

Figure 12-10   Input RC compensation.

TABLE 12-1

	Oscillates unloaded?			
		Oscillates with V_{in} = 0 V?		
			Loop check[#] fixes oscillation?	
				Probable cause in order of probability (Table 12-2)
CLBW $\leq f_C \leq$ UGBW	N	Y	N	A, C, D, B
CLBW $\leq f_C \leq$ UGBW	Y	Y	Y	J, E, F, I
CLBW $\leq f_C \leq$ UGBW	–	N	Y	G
$f_C \leq$ UGBW	N	Y	Y	D
f_C = UGBW	Y	Y	N*	I, C
$f_C \ll$ UGBW	Y	Y	N	K, C
$f_C >$ UGBW	N	Y	N	B, A
$f_C >$ UGBW	N	N**	N	A, B, H

f_c is the frequency of oscillation
CLBW is the closed-loop bandwidth; bandwidth with the selected closed-loop gain A_V
UGBW is the unity gain-bandwidth; bandwidth with $A_V = 1$
[#]See Fig. 12-11 for loop check circuit
⁻M or may not make a difference; indeterminate
*Loop check (12-11) stops oscillation if $R_n \ll |A_{in(eff)}|$
**Oscillates only over a portion of the output cycle
(Courtesy of Apex Microtechnology Corporation)

Tables 12-1 and 12-2 show one power Op Amp manufacturer's list of oscillating frequencies, their probable causes, and possible solutions.

Recall that back in Chapter 6, Section 6.2, we learned that, with low ratios of R_F/R_1, oscillations can occur. That is, an Op Amp circuit designed to have less gain will inherently have more feedback. This feedback shifts to a (more or less) in-phase condition at higher frequencies and can become positive (regenerative). Positive feedback makes any amplifier less stable.

If an amplifier circuit is oscillating, a *loop check* can be performed to determine if the cause is excessive positive feedback. As shown in Fig. 12-11, this simply involves raising the circuit gain, with the shunt resistor R_n, to see if the oscillations stop. If they do, probable causes and possible solutions are discussed in Table 12-1.

12.7 ISOLATING A CAPACITIVE LOAD

As shown in Fig. 12-12, inductance L_0 in the range of 3–10 μH can be placed in the output lead if the load is capacitive. A parallel resistor R_0 in

TABLE 12-2

A	Cause:	Feedback through the dc supply (insufficient bypassing)
	Solution:	Properly bypass the dc supplies; see Fig. 12-9
B	Cause:	Supply lead inductance (insufficient bypassing)
	Solution:	Properly bypass the dc supplies; see Fig. 12-9
C	Cause:	Ground loops
	Solution:	Use star grounding; see Fig. 12-9
D	Cause:	Capacitive load reacting with output impedance
	Solution:	Increase closed-loop gain or use RC input compensation; see Fig. 12-10
E	Cause:	Inductance in the feedback loop
	Solution:	Use an alternate feedback path
F	Cause:	Input capacitance reacting with R_F
	Solution:	Use C_F in parallel with R_F; too much C_F can cause Problem I
G	Cause:	Coupling of output to input
	Solution:	Run output PCB traces away from input traces, bypass or eliminate R_2 (the bias current compensating component between the noninverting input and ground).
H	Cause:	Emitter follower output stage reacting with capacitive load
	Solution:	Use an output snubber; see Section 12.8
I	Cause:	Feedback capacitor C_F on amplifier that is not unity gain stable; integrator instability
	Solution:	Use a smaller C_F and/or a larger compensating capacitor C_C; see the TYPICAL APPLICATION in the Specs of the PB58 in the Appendix T.
J	Cause:	Insufficient compensating capacitance C_C for the closed-loop gain used
	Solution:	Increase compensating capacitor C_C or gain of use input compensation; see Specs of the PB58 and Fig. 12-10
K,	Cause:	Servo loop stability problem
	Solution:	Compensate the servo amplifier

(Courtesy of Apex Microtechnology Corporation)

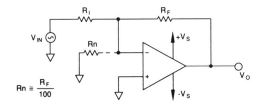

Figure 12-11   Loop check circuit.

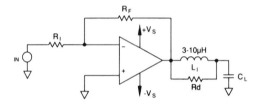

Figure 12-12   Isolating a capacitive load.

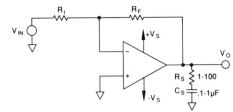

Figure 12-13   Output RC snubber for inductive loads.

the range of 5–100 Ω serves to suppress resonance. This $L_0 R_0$ combination in series with the capacitive load makes the output of the Op Amp essentially "see" a resistive load.

12.8   AN OUTPUT SNUBBER

Snubbing refers to suppression of high-frequency oscillations caused by inductive loads. A snubber consists of series components R_S and C_S across the inductive load as shown in Fig. 12-13. The impedance of this snubber is low at high frequencies and thus helps to keep these frequencies out of the local feedback loop. Where a snubber is required, R_S is typically in the range of 10–100 Ω and C_S is in the range 0.1–1 μF.

12.9   SLEEP MODE

For battery-powered applications, some power Op Amps have a *sleep mode* feature that can lower the quiescent current at times when the input signal is zero. The WB05 in Fig. 12-14 is such an amplifier. This amplifier's quiescent current I_Q decreases from about 30 to 2.5 mA when both sleep pins are pulled to within 0.1 V of their respective rails.

Referring to the *equivalent schematic* of Fig. 12-14, note that if the sleep pins are left open, current will flow though the zener diodes $D1$ and $D2$ via the resistor below $D1$. The voltage drop across $D1$ drives $Q1$ into saturation. Similarly, the drop across $D2$ saturates $Q17$. This admits current through

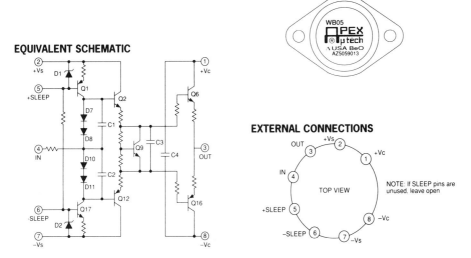

Figure 12-14 Power Op Amp with a sleep mode feature.

$D7$, $D8$, $D10$, and $D11$. The drop across these diodes causes drivers $Q2$ and $Q12$ to conduct current. This current through their emitter resistors places enough forward bias on the base-emitter junctions of $Q6$ and $Q16$ that they conduct a quiescent current I_Q of about 30 mA. This partial conduction serves to reduce *crossover* distortion.

For simplicity, the circuit of Fig. 12-1 was used to describe the action of a *push-pull class B* power amplifier. However, it lacks some practical features. As is, this circuit of Fig. 12-1 introduces crossover distortion because its output BJTs are at cutoff under quiescent conditions. Signal V_{out} must exceed about ± 0.65 V before V_o moves from 0 V. When input signal is applied, this causes V_o to be a distorted version of V_{out}, especially with small amplitudes of V_{out}. In the circuit of Fig. 12-14, on the other hand, $Q2$ and $Q12$ are already partially on under quiescent conditions because of the voltage drops across diodes $D7$, $D8$, $D10$, and $D11$. Therefore, *any* changes in the input signal on pin 4 will result in proportional changes in conductivity of these power BJTs. Therefore, compared to the circuit of Fig. 12-1, the signal at the output (pin 3 of Fig. 12-14) is a much more linear function of the input signal.

If the SLEEP + pin is placed at or near the $+V_S$ potential and the SLEEP − pin is at or near $-V_S$, the voltage across both zeners drops below 0.65 V. This causes the complementary BJTs $Q1$ and $Q17$ to become nonconducting, which in turn cause $Q2$ and $Q12$ to cut off too. This reduces the base-emitter voltages on the power output BJTs, causing their quiescent current to drop to about 2.5 mA.

The circuit of Fig. 12-15 shows how the sleep mode can be controlled by a digital control signal ($\overline{\text{SLEEP}}$). When $\overline{\text{SLEEP}}$ is LOW, all three BJTs conduct

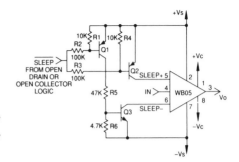

Figure 12-15 Sleep mode controlled by a digital input signal $\overline{\text{SLEEP}}$. Courtesy of Apex Microtechnology Corporation.

(saturate). Saturated $Q2$ and $Q3$ cause the SLEEP pins to go to their respective rails and this puts the WB05 into its sleep mode. On the other hand, if signal $\overline{\text{SLEEP}}$ is HIGH, all three external BJTs cut off. Both SLEEP pins then float, causing this amplifier to come out of its sleep mode.

12.10 SELECTING A HEAT SINK

An appropriate heat sink keeps the junction temperature relatively low. This adds to the reliability of the power Op Amp. Figure 12-16 shows that the failure rate of a silicon junction rises exponentially with temperature. Note that the failure rate at 175° is almost 3 times that of 125°. Also, at higher temperatures, the input offset voltages and bias currents drift in power Op Amps. Dynamic characteristics, slew rates, open-loop gains, bandwidths, etc., are affected too.

Figure 12-17 shows an electrical equivalent of a power Op Amp and its heat sink. The amount of power, in the form of heat, that flows from a power transistors' junctions to the surrounding air can be expressed with an equation similar to Ohm's law:

$$P_D = \frac{T_J - T_A}{\theta_{JC} + \theta_{CS} + \theta_{SA}}, \qquad (8\text{-}12)$$

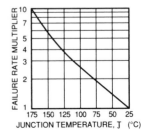

Figure 12-16   JUNCTION TEMPERATURE, J (°C)

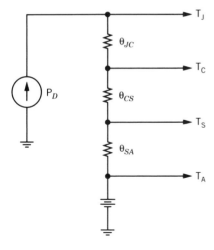

Figure 12-17 Thermoelectric model.

where P_D is the power dissipated at the junctions in watts (W)

T_J is the temperature of the junctions in degrees Celsius (°C)

T_A is the ambient (surrounding air) temperature in °C

θ_{JC} is the junction-to-case thermal resistance in °C/W, which is specified by the power Op Amp manufacturer

θ_{CS} is the case-to-sink thermal resistance in °C/W, which is typically in the range of 0.1–0.2°C/W with good mounting practices including thermally conducting grease

θ_{SA} is the sink-to-air thermal resistance in °C/W, which is specified by the heat sink manufacturer

Thus, when selecting a heat sink, we must know the maximum power that the power Op Amp is required to dissipate ($P_{D(\text{max})}$) and the highest expected ambient temperature (T_A) and maximum junction temperature (T_J) that will provide acceptable performance and reliability. If its value is significant, it does not hurt to add the quiescent power (P_Q) to the maximum dissipated power caused by input signal.

Example 12-4

Suppose that we have a power Op Amp that must be capable of dissipating a $P_{D(\text{max})}$ of 10 W, as was determined by Eq. (12-4a). Assume that this is within the SOA of this Op Amp. Its spec sheet specifies a $\theta_{JC} = 4°C/W$. With thermally conducting grease between the case and the sink, we can expect $\theta_{CS} = 0.2°C/W$. If we can let the junction temperature be as high as 125°C and rely on an ambient temperature that is no higher than 35°C, (a) what maximum case-to-air thermal resistance can its heat sink have assuming that the quiescent power dissipation is negligible? If (b), the quiescent power dissipation is 0.3 W and we include it for a safety margin, what maximum case-to-air thermal resistance can the heat sink have?

Answer. Rearranging Eq. (8-12), we can show that the heat sink's maximum thermal resistance

$$\theta_{SA} < \frac{T_J - T_A}{P_{D(max)}} - \theta_{JC} - \theta_{CS}.$$

(a) With a negligible P_{DQ}, we would choose a heat sink with a thermal resistance

$$\theta_{SA} < \frac{125° - 35°}{10 \text{ W}} - 0.2° \text{ C/W} - 4° \text{ C/W} = 4.8° \text{ C/W}.$$

(b) Including a 0.3-W quiescent power, we would show that

$$P_{D(max)} = 10 \text{ W} + 0.3 \text{ W} = 10.3 \text{ W},$$

and therefore,

$$\theta_{SA} < \frac{125° - 35°}{10.3 \text{ W}} - 0.2° \text{ C/W} - 4° \text{ C/W} = 4.54° \text{ C/W}.$$

12.11   MOUNTING CONSIDERATIONS

Power Op Amps are commonly destroyed in the application development stage while mounting them on a heat sink. Excessive torque on the mounting bolts will flex the case and crack the substrate. It is important, therefore, to carefully follow the manufacturer's instructions, recommended torque, type of washers, etc. Another frequent cause of power Op Amp failure is insufficient heat sinking.

When a power transistor is dissipating power, the heat flow spreads downwards as it travels through various materials on its way to the heat sink; see Fig. 12-18a. From there it flows past the drilled pin holes into the bulk of the material of the sink; see Fig. 12-18b. The sink surface directly below the junction, therefore, conducts the greatest concentration of heat and must not be drilled out in place of individual small pin holes.

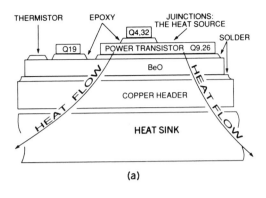

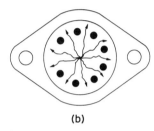

Figure 12-18 (a) Heat flow path from junction to heat sink. (b) Path of heat in the sink material directly below the case. (Courtesy of Apex Microtechnology Corporation.)

Various mounting techniques are shown in Fig. 12-19. In these examples, the Op Amps have electrically isolated cases. These cases, therefore, can be directly mounted to the sink for better heat conductivity. Depending on the manufacturer, some Op Amp types require an electrically insulating washer between the case and the sink. Of course, this adds to the thermal resistance between the case and the sink, which can affect the size of the heat sink that must be used. With either type, Teflon tubes must be used on the pins to prevent them from shorting through the sink.

12.12 MOTOR DRIVES

Motor control applications usually place brutal demands on the driving circuit. The mechanical inertia of the rotating motor and its load can cause stress on the output transistors of the power amplifier. The motor in Fig. 12-20a is electrically equivalent to an inductance L and resistance $R_{(eff)}$ in series. Resistance $R_{(eff)}$ is the *effective resistance* of the motor, 4 Ω in this example. It is not simply the resistance of the motor windings. V_o is the sum of the drops across L and $R_{(eff)}$, shown here as a steady-state 26 V.

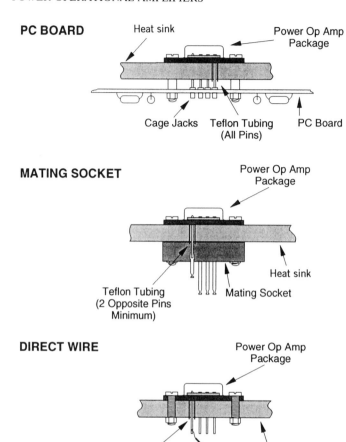

PC BOARD

Heat sink

Power Op Amp Package

Cage Jacks   Teflon Tubing   PC Board
              (All Pins)

MATING SOCKET

Power Op Amp Package

Heat sink

Teflon Tubing   Mating Socket
(2 Opposite Pins
Minimum)

DIRECT WIRE

Power Op Amp Package

Teflon Tubing   Heat sink
(All Pins)
              Wire

Figure 12-19 Mounting techniques. (Courtesy of Apex Microtechnology Corporation.)

Figure 12-20b shows an equivalent circuit when Q_1 is conducting, Q_2 cut off, and the motor drawing 3 A. The rotating armature has an induced electromotive force (emf) shown here as a dc equivalent voltage source. At steady state, that is, when the motor is delivering constant mechanical power to its load, the emf opposes the dc supply voltage V_S. The value of emf is a function of the motor speed; generally, more or less speed *generates* more or less emf. Since this motor draws $I_o = 3$ A, we can show that

$$\text{emf} = V_o - R_{(\text{eff})}I_o = 26 \text{ V} - (4 \text{ }\Omega)3 \text{ A} = 14 \text{ V}.$$

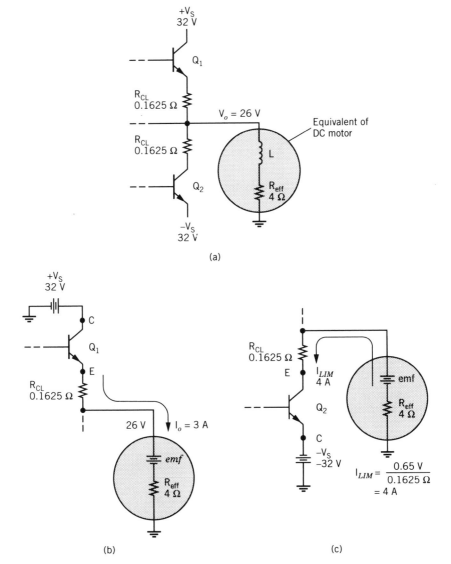

Figure 12-20   (a) Power stage driving a motor. (b) Motor draws a steady-state 3 A. (c) Electrical equivalent at instant when Q_1 turns off and Q_2 turns on

Generally, Q_1's collector-to-emitter voltage

$$V_{CE_1} = V_S - R_{LIM}I_o - V_o.$$

In this case, therefore,

$$V_{CE_1} = 32 \text{ V} - (0.1625 \text{ }\Omega)3 \text{ A} - 26 \text{ V} = 5.51 \text{ V}.$$

And Q_1's power dissipation

$$P_{Q_1} = V_{CE_1}I_o = 5.51 \text{ V}(3 \text{ A}) \cong 16.53 \text{ W}.$$

With a reversal command, Q_1 quickly switches to nonconducting and Q_2 to conducting states, and the equivalent circuit becomes Fig. 12-20c. When these BJTs switch, the mechanical inertia of the motor and its load cause emf to remain the same, at least at that instant. We will assume that Q_2 has turned on hard enough to admit a sink current of 4 A; that is, solving Eq. (12-6) for I_{LIM},

$$I_{LIM} = \frac{0.65 \text{ V}}{R_{CL}} = \frac{0.65 \text{ V}}{0.1625 \text{ }\Omega} = 4 \text{ A}.$$

This equivalent circuit shows that at the instant of the reversal command, the rail voltage $-V_S$ and emf are in series aiding and that Q_2's collector-to-emitter voltage

$$V_{CE_2} = +V_S - R_{\text{eff}}I_{LIM} + \text{emf} - R_{CL}I_{LIM}$$

or

$$V_{CE_2} = +32 \text{ V} - (4 \text{ }\Omega)4 \text{ A} + 12 \text{ V} - (0.1625 \text{ }\Omega)4 \text{ A} = +27.35 \text{ V}.$$

And therefore, Q_2's power dissipation

$$P_{Q_2} = V_{CE_2}I_{LIM} = 27.35 \text{ V}(4 \text{ A}) \cong 109.4 \text{ W}.$$

We see that the mechanical energy stored in the rotating motor and its load, combined with the power from the $-V_S$ supply, can cause momentary power dissipation that is much greater than under steady-state conditions; more than six times greater in this example. The action here is like that in the time-seasoned practice of *dynamic breaking*. That is, in industry, motors are commonly slowed or stopped by switching power resistors across them to absorb kinetic energy.

Use of power Op Amps with power dissipating capabilities much higher than needed to support steady-state running conditions is usually not practical. Instead, the higher transient power dissipations can be reduced by limiting the rate of change of the drive amplifier's output voltage to approximately match the rate of change imposed by the mechanical inertia of the motor and its load.

REVIEW QUESTIONS 12

12-1. Name two advantages of a single package power amplifier over a discrete part version.

12-2. What is meant by the term *complementary pair*?

12-3. When a power Op Amp, or any electronic device, is said to be either sourcing or sinking current, what do these terms mean?

12-4. Describe class B operation. Also, class A operation.

12-5. What is a power amplifier's quiescent current? What is its quiescent power?

12.6. A power Op Amp amplifier with a given load resistance will be used sometimes with ac signals, sometimes with dc inputs. With which would you expect greater possible heat at the junctions of the individual power output transistors?

12.7. What does the abbreviation SOA mean?

12.8. Explain what is meant by thermal runaway of a BJT.

12.9. How is the output current of a power Op Amp prevented from rising to a destructive value if, say, the output is accidentally shorted to ground?

12-10. What is inductive kickback? What is done to reduce its adverse effects?

12-11. Name three causes of instability of a power Op Amp.

12-12. Name five design techniques used to minimize the possibility of unstable operation.

12-13. If a power Op Amp has a sleep mode feature, what operating characteristics does it offer?

12-14. Suppose that for cost- and space-saving advantages, you are asked to replace a power Op Amp's given heat sink with a much smaller one. What adverse effects would you predict?

12-15. What do each of the following terms mean?: T_J, T_A, θ_{JC}, θ_{CS}, and θ_{SA}.

12-16. For efficient heat flow from a BJT's junction to the surrounding air, we want as (*high*), (*low*) as possible thermal resistances in between.

12-17. For efficient heat flow from a BJT's junction to the surrounding air, we want as (*high*), (*low*) as possible temperature *difference* between the case and air.

12-18. To get good contact and therefore efficient heat flow from the case to the sink, the case should be bolted down very firmly to the case (*True*, *False*). Explain why.

12-19. If the case of a power Op Amp is electrically isolated, should we use an insulating washer between the case and sink just to be safe? Why?

12-20. The production manager says that he can save the company a lot of money if he can drill one large hole in the heat sink instead of eight small holes. How would you answer?

12-21. If a power Op Amp is to be used to drive a motor for variable speeds in both directions, what precautions would you take or recommend?

PROBLEMS 12

12-1. A power Op Amp is required to delivered sinusoidal power at a broad range of audio frequencies to its load. The rail voltages are ± 32 V and the load is 15 Ω. Find the maximum power that this amplifier must be capable of dissipating.

12-2. A power Op Amp is required to deliver dc power, with positive and negative voltages, to a 15-Ω load. The rail voltages are ± 32 V. Find the maximum power that this amplifier must be capable of dissipating.

12-3. Referring to the amplifier described in Problem 12-1, if its quiescent current is specified at 4 mA, how much quiescent power does it dissipate?

12-4. Referring to the amplifier described in Problem 12-2, if its quiescent current is specified at 2.5 mA, how much quiescent power does it dissipate?

12-5. Referring to the current described in Problem 12-1, what maximum thermal resistance can the heat sink have if we need to keep the junction temperature below 120°C and the ambient temperature is likely to be as high as 45°C? The junction-to-case thermal resistance is 3.8°C/W and the case-to-sink thermal resistance is 0.15°C/W. Ignore its quiescent power dissipation.

12-6. Referring to the circuit described in Problem 12-2, what maximum thermal resistance can the heat sink have if we need to keep the junction temperature below 120°C and the ambient temperature is likely to be as high as 45°C? The junction-to-case thermal resistance is 3.8°C/W and the case-to-sink thermal resistance is 0.15°C/W. Ignore its quiescent power dissipation.

12-7. Select the current limiting resistors for the power Op Amp of Fig. 12-7 to limit the current to 1.5 A. Across which pins would you connect them?

12-8. What is the current limit of the current of Fig. 12-7 if the current limiting resistors are each 0.5 Ω?

GLOSSARY

Amplification, Voltage (A_v) The closed-loop voltage gain of an Op Amp, the voltage gain with negative feedback.

Amplification, Differential (A_d) The ratio of output voltage to the differential input voltage V_{id} of a differential amplifier.

Amplification, Voltage Open-Loop (A_{VOL}) The ratio of the output voltage V_o to the differential input voltage V_{id} of an Op Amp.

Average Temperature Coefficient of Input Offset Current The operating temperature range divided into the resulting change of the input offset current ΔI_{io}.

Average Temperature Coefficient of Input Offset Voltage The operating temperature range divided into the resulting change of the input offset voltage ΔV_{io}.

Bandwidth (BW) The frequency range in which the Op Amp's gain does not drop by more than 0.707 or 3 dB of its dc value.

Beta (β) That ratio of a bipolar transistor's (BJT's) collector current to its base current; also called h_{fe}.

BJT Bipolar junction transistor. Standard PNP or NPN type transistors, often called simply junction transistors.

Bootstrapping Technique of using positive feedback to raise an amplifier's input impedance or resistance.

Buffering Use of an isolating amplifier that prevents variation in load resistance from affecting the source of signal driving it.

Channel separation The ratio of the output signal voltage of a driven amplifier to the output signal voltage of an adjacent undriven amplifier on the same chip; usually expressed in decibels.

Chip A small piece of semiconductor (monolith) in or on which an integrated circuit is built.

Chopper Stabilized Op Amp A high-performance Op Amp, usually a hybrid, having a very small input bias current and low drift.

CMR (dB) See Common-Mode Rejection Ratio.

Common-Mode Input Resistance The resistance with respect to ground or a common point looking into both inputs tied together.

Common-Mode Rejection Ratio (CMRR) The ratio of the closed-loop gain A_v to the common-mode gain A_{cm}; also the ratio of the change in the common-mode input voltage ΔV_{cm} to the resulting change in the input offset voltage ΔV_{io}. Called CMR (dB) when expressed in decibels.

Common-Mode Input Voltage Swing The peak value of common-mode input voltage that can be applied and still maintain linear operation.

Common-Mode Input Voltage (V_{cm}) Input voltage, usually noise, that appears at both inputs of a differential or operational amplifier simultaneously.

Common-Mode Output Voltage The output voltage resulting from a voltage applied to both inputs simultaneously.

Common-Mode Gain (A_{cm}) The ratio of common-mode output voltage to the common-mode input voltage V_{id} of a differential or operational amplifier.

Compensated Term used to describe Op Amps that are internally compensated and have no provisions for nor require external compensation; also called internally compensated.

Compensation Use of resistors and/or capacitors to stabilize an Op Amp circuit; alter its A_{VOL} vs. frequency characteristics.

Complementary Pair A PNP and an NPN transistor working in a power amplifier stage.

DAC Digital-to-analog converter. See D/A in the index.

Darington Pair Two low-leakage bipolar transistors wired so that their total beta (β), or h_{fe}, is the product of their individual β's.

DC Power Dissipation See Internal Power Dissipation.

Drift Changes in component or circuit parameters caused by changes in temperature, supply voltage, or time.

Drive Application of signal to the input of an amplifier.

Dual Supply More than one dc power supply voltage, usually opposite polarities. See Split Supply.

Effective Input Resistance $(R_{i(\text{eff})})$ Small-signal ac input resistance seen looking into the appropriate input with the other input grounded or common under closed-loop conditions.

Effective Output Resistance ($R_{o(\text{eff})}$)   Small-signal ac output resistance looking back into the output of an Op Amp under closed-loop conditions.

Fall Time   The time required for an output voltage to change from 90% to 10% of its final value in response to a step input voltage.

Feedback   Use of circuitry that applies a fraction of the output signal back to one of the inputs of an Op Amp. See Positive Feedback and Negative Feedback.

FET   Field-effect transistor. A generic term. Also called unipolar transistors. See BJT and MOSFET.

Gain-Bandwidth Product (A_vBW)   The product of a compensated Op Amp's closed-loop gain and its bandwidth with that gain.

Hybrid Op Amp   Op Amp that contains ICs and discrete components.

Input Offset Current Drift (ΔI_{io})   The changes in the input offset current with changes in temperature, supply voltage, or time.

Input Bias Current (I_B)   The average of the two dc input bias currents measured while both inputs are grounded or connected to a common point.

Input Offset Current (I_{io})   The difference in the two dc input bias currents measured while both inputs are grounded or connected to a common point.

Input Voltage Range   The range of voltage that can be applied to either input over which the Op Amp is linear.

Input Bias Current Drift (ΔI_B)   The changes in the input bias current with changes in temperature, supply voltage, or time.

Input Offset Voltage (V_{io})   The voltage that must be applied across the inputs to force the output to zero volts under open-loop conditions.

Input Resistance (R_i)   Small-signal ac input resistance seen looking into the appropriate input with the other input grounded or common under open-loop conditions.

Internal Power Dissipation   The power required to operate the amplifier with an open load and no input signal; also called dc power dissipation. See Quiescent Power.

Internally Compensated   See Compensated.

Latch-Up   A condition where the output of an Op Amp hangs up on either the positive or negative rail.

Loop Gain   The ratio of the open-loop gain A_{VOL} to the closed-loop gain A_v.

Monolithic IC   An integrated circuit with components formed on or within a single piece of semiconductor called a substrate.

MOSFET   Metal oxide semiconductor field-effect transistor; also called IGFET (insulated gate field-effect transistor).

Negative Feedback   Use of circuitry that applies a fraction of the output signal back to the inverting input of an Op Amp.

Negative Rail   The negative dc supply voltage.

Noise Margin (NM)   The difference between the upper and lower threshold voltages of a window comparator; that is, a comparator with hysteresis in its transfer function.

Noise Figure (NF)   The ratio of an Op Amp's input signal-to-noise ratio to its output signal-to-noise ratio expressed in decibels.

Open-Loop Gain (A_{VOL})   The ratio of the output voltage V_o to the differential input voltage V_{id} of an Op Amp.

Oscillator   A circuit that is capable of producing electrical oscillations. Typical oscillator outputs are sine waves, square waves, sawtooth waves, triangular waves, etc.

Output Short-Circuit Current   The output current with the output shorted to ground or common.

Output Offset Voltage (V_{oo})   The dc output of an Op Amp before its input is nulled; a textbook term used here for pedagogical purposes.

Output Voltage Swing   The range over which the output signal voltage can vary without clipping.

Output Resistance (R_o)   Small-signal ac output resistance looking back into the output of an Op Amp under open-loop conditions.

Overshoot   The output voltage swing beyond its final quiescent value in response to a step input voltage.

Positive Feedback   Use of circuitry that applies a fraction of the output signal back to the noninverting input of an Op Amp.

Positive Rail   The positive dc supply voltage.

Power Supply Sensitivity (PSS)   The ratio of the change in the input offset voltage to the change in the power supply voltage causing it. The reciprocal of power supply rejection ratio.

Power Supply Rejection Ratio (PSRR)   The ratio of the change in the power supply voltage to the resulting change in the input offset voltage; usually expressed in decibels. The reciprocal of the power supply sensitivity factor.

Power Supply Current   See Quiescent Current.

Quiescent Current (I_Q)   The current the Op Amp draws from the dc rails with the load open and zero output voltage.

Quiescent Power   The power consumed by an Op Amp with no load and its output at zero volts.

Rail   See Positive Rail or Negative Rail.

Rate of Closure   The rate at which the open-loop gain vs. frequency curve is decreasing as it intersects the horizontal closed-loop gain projection; usually expressed in decibels per decade.

Ringing   Oscillations of the output voltage about the final quiescent value in response to a step input voltage.

Ripple Rejection   See Power Supply Rejection Ratio (PSRR).

Rise Time   The time required for an output voltage to change from 10% to 90% of its final value in response to a step input voltage.

Roll-Off   The decrease in an amplifier's gain at higher frequencies.

RTI (Referred To Input)   Errors such as input offset voltage V_{io} are assumed to be applied to the noninverting input.

Safe Operating Area   See SOA.

Secondary Breakdown   Caused by the tendancy of *current crowding* to occur in a bipolar transistor at higher V_{CE} voltages. Current crowding causes localized hot spots within the semiconductor and requires derating of its maximum power dissipation.

Setting Time   The time between the instant a step input voltage is applied and the time the output settles to a specified percentage of the final quiescent value.

Sleep Mode   Mode of operation available in some types of Power Op Amps in which its power dissipation is reduced during quiescent conditions.

Slew Rate (SR)   The minimum rate of change of the output voltage under large-signal conditions; usually listed in volts per microsecond.

SOA   Safe operating area. Indicates power dissipating limits on an Op Amp's current vs voltage characteristics.

Split Supply   Two dc power supply voltages, equal but opposite polarities. Also called a *balanced* power supply.

Standby Current Drain   The part of the output current of a power supply that does not contribute to the load current; also see Quiescent Current.

Substrate   Base semiconductor material on which an integrated circuit is built.

Supply Current   The current drained from the power supply to operate the amplifier with no load and the output voltage halfway between the dc supply voltages; also see Quiescent Current.

Tailored Response   Selecting an Op Amp's A_{VOL} vs. frequency characteristics by choosing appropriate external compensating components; also see Compensation.

Transfer Function   The output voltage vs. input voltage curve.

Uncompensated   Op Amps that are not internally compensated, often referred to as having a tailored response.

Unity Gain Bandwidth   Bandwidth when the Op Amp is wired to have a closed-loop gain of 1.

Zener Knee Current   The minimum zener current required to keep the zener diode in zener (avalanche) conduction.

Zener Test Current   The zener current with which the specified zener voltage and zener resistance is measured.

APPENDIX A

DERIVATION OF EQ. (1-6)

Referring to Fig. A-1, the resistance to ground looking up through R_E includes R_E and parallel paths across the emitter-base junctions and the R_B resistances. These R_Bs are typically the internal resistances of the signal voltage sources V_1 and V_2 and are smaller by the factor h_{fe} as viewed up form R_E. In other words, because base current is smaller than the emitter current by this factor h_{fe}, the effective value of resistance in each base must be smaller by this same factor from the emitter's point of view. Thus the total resistance is

$$R_E + R_B/2h_{fe},$$

assuming that each base has the same value of R_B.

Also assuming that each base-emitter junction has the same voltage drop, the total voltage across the total resistance is

$$V_{EE} - V_{BE}.$$

Therefore, the current

$$I_E = \frac{V_{EE} - V_{BE}}{R_E + R_B/2h_{fe}}. \tag{1-6}$$

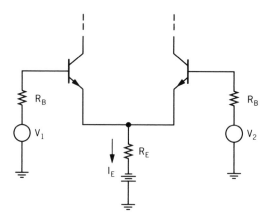

Figure A-1   Elementary current source.

APPENDIX B

DERIVATION OF EQS. (1-12a) AND (1-12b)

The circuit of Fig. B-1 is the equivalent of the circuit of Fig. 1-5 as the signal sources V_1 and V_2 see it. The resistance between points B_1 and E is the dynamic (ac) base-to-emitter resistance of transistor Q_1. Similarly, the resistance between points B_2 and E is the dynamic (ac) base-to-emitter resistance of transistor Q_2. Current i_b "sees" resistance r'_e larger by the factor h_{fe} because i_b is smaller than i_c by this factor. By Ohm's law we can show that

$$V_1 + V_2 = i_b(2h_{fe}r'_e + 2R_B).$$

Factoring 2 from each term yields

$$V_1 + V_2 = 2i_b(h_{fe}r'_e + R_B).$$

And since $i_b = i_c/h_{fe}$, the above becomes

$$V_1 + V_2 = 2i_c(r'_e + R_B/h_{fe}). \tag{B-1}$$

When a BJT differential amplifier is driven with an input signal ($V_{id} > 0$), a base current i_b flows. It causes each BJT to output a collector current i_c. Figure B-2 shows an equivalent of the circuit of Fig. 1-5 as the BJTs Q_1 and Q_2 "see" it. Note that on alternations of V_{id} that cause current through Q_1 to increase (an upward current i_c), the current through Q_2 decreases (i_c is downward). On the opposite alternations, then, Q_1 conducts downward and Q_2 conducts upward.

Note in Fig. B-2 that the collector signal current i_c "sees" resistors R_{C1} and R_{C2} in series. The sum of the voltages across them is the differential

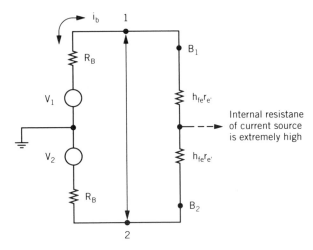

Figure B-1   Equivalent circuit as signal sources V_1 and V_2 see it.

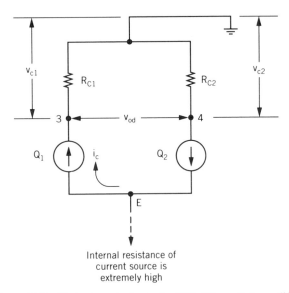

Figure B-2   Output equivalent of a BJT differential amplifier.

output signal V_{od}. By Ohm's law, this voltage can be shown

$$V_{od} = (R_{C1} + R_{C2})i_c.$$

And if R_{C1} and R_{C2} are equal, say equal to R_C, the above equation becomes

$$V_{od} = 2R_C i_c. \tag{B-2}$$

Since the differential voltage gain

$$A_d = \frac{V_{od}}{V_1 + V_2},$$

we can substitute Eqs. (B-1) and (B-2) into the above and get

$$A_d = \frac{2R_C i_c}{2i_c(r'_e + R_B/h_{fe})}.$$

This simplifies to

$$A_d = \frac{R_C}{r'_e + R_B/h_{fe}}. \tag{1-12a}$$

And if $R_B = 0$, then

$$A_d = \frac{R_C}{r'_e}. \tag{1-12b}$$

APPENDIX C

DERIVATION OF EQS. (1-13) AND (1-18): DYNAMIC RESISTANCE OF SILICON DIODE JUNCTION

The dynamic resistance r_e' seen looking into the emitter of a forward-biased emitter-base (diode) junction can be found with the equation

$$r_e' = \frac{kT/q}{I_E},$$

where k is Boltzmann's constant

$$(1.38 \times 10^{-23} \text{ J/K})$$

T is the absolute temperature kelvin

q is the charge of electron

$$= 1.602 \times 10^{-19} \text{ coulomb}$$

I_E is the emitter or collector current, amperes

The absolute temperature T is expressed in degrees kelvin, which is 273° plus the Celsius temperature. Therefore, room temperature of 25°C is 298°K. If the junction is at room temperature, the dynamic resistance is about

$$r_e' = \frac{kT/q}{I_E} \simeq \frac{30 \text{ mV}}{I_E}.$$

At normal temperatures, this dynamic resistance is in the range

$$\frac{25 \text{ mV}}{I_E} \leq r_e' \leq \frac{50 \text{ mV}}{I_E}. \tag{1-13}$$

APPENDIX D

DERIVATION OF EQ. (1-19)

An input equivalent circuit of a JFET differential amplifier is shown in Fig. D-1a. The differential input voltage

$$V_{id} = V_1 - V_2.$$

Since the gate-to-source junctions of these JFETs are reverse biased, the resistance looking into each gate is extremely high and the internal resistances of signal sources of V_1 and V_2 can be considered negligible for most practical purposes and are not shown. Assuming that these JFETs are well matched, half of V_{id} is dropped across each gate-source junction; see Fig. D-1b.

The transconductance g_m of a FET is defined by the equation

$$g_m \approx \Delta I_D / \Delta V_{GS}$$

or

$$g_m \approx i_d / v_{gs},$$

where v_{gs} is the signal applied to the gate-source junction ($V_{id}/2$ in Fig. D-1) and i_d is the resulting signal current in the drain. Rearranging the latter, we show that

$$v_{gs} = v_{id}/2 \approx i_d / g_m$$

or

$$V_{id} \approx 2 i_d / g_m. \tag{D-1}$$

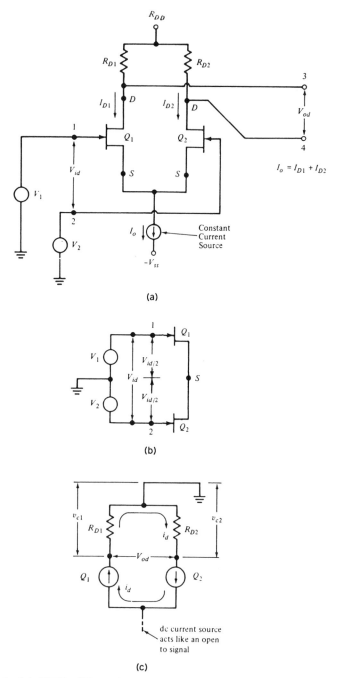

Figure D-1 (a) JFET differential amplifier circuit; arrows indicate conventional current direction. (b) Input equivalent and (c) gate-source junctions act like voltage-divider resistors. (d) ac equivalent of output.

The circuit of Fig. D-1c shows the output equivalent circuit of a JFET differential amplifier. When it is driven with an input signal ($V_{id} > 0$), a drain current i_d flows. Note that on alternations of V_{id} that cause current through Q_1 to increase (an upward current i_d), the current through Q_2 decreases (i_d is downward). On the opposite alternations, then, Q_1 conducts downward and Q_2 conducts upward.

Signal current i_d "sees" resistors R_{D1} and R_{D2} in series. The sum of the voltages across them is the differential output signal V_{od}. By Ohm's law, this voltage can be shown

$$v_{od} = (R_{D1} + R_{D2})i_d.$$

If we let $R_D = R_{D1} = R_{D2}$, the above becomes

$$v_{od} = 2R_D i_d. \tag{D-2}$$

The JFET circuit's differential gain

$$A_d = \frac{v_{od}}{v_{id}}.$$

Substituting Eqs. (D-1) and (D-2) into this yields

$$A_d = \frac{2R_D i_d}{2i_d/g_m} = g_m R_D. \tag{1-19}$$

The signal voltage from either drain with respect to ground is equal to the signal voltage across either drain resistor. Therefore,

$$|v_{d1}| = |v_{d2}| = v_d = R_D i_d,$$

where v_{d1} and v_{d2} are equal in amplitude but $180°$ out of phase. The ratio of the signal voltage at either drain to the differential input voltage is one-half of the differential gain A_D; that is,

$$A_v = \frac{v_d}{v_{id}} = \frac{g_m R_D}{2}.$$

This is true because the differential output voltage v_{od} is the sum of the voltages across resistors R_{D1} and R_{D2}, whereas v_d is the signal voltage across only one of these resistors.

APPENDIX E

TABLE OF OP AMPS AND SPECIFICATIONS

TABLE E-1. Linear Operational Amplifiers $T_A = 25°C$.

Device	Operating Temp. Range °C Min	Max	$A_{VOL.}$ Min K	R_i Min Ω	P_D mW	$\|I_{io}\|$ Max nA	I_b Max nA	CMV_i Min V	*Typ. Slew Rate SR V/μs	V_o Min V	$\|V_{io}\|$ Max mV	Offset Adjust	Internal Compensation	Output Protection	Input Protection	JEDEC Package Type
709A	−55	+125	25	350 k	108	50	200	±8	0.3	±12	1	no	no	no	no	TO-91, 99, 116
709B	−55	+125	25	150 k	165	200	500	±8	0.3	±12	5	no	no	no	no	TO-91, 99, 116
709C	0	+70	15	50 k	200	500	1500	±8	0.3	±12	10	no	no	no	no	TO-91, 99, 116
739C(Dual)	0	+70	6.5	37 k	420	1000	2000	±10	1.0	±12, −14	6	no	no	yes	yes	TO-116
741B	−55	+125	50	300 k	85	200	500	±12	0.5	±12	5	yes	yes	yes	yes	TO-91, 99, 116
741C	0	+70	20	150 k	85	200	500	±12	0.5	±12	6	yes	yes	yes	yes	TO-91, 99, 116
747B(Dual)	−55	+125	50	300 k	85	200	500	±12	0.5	±12	5	yes	yes	yes	yes	TO-101, 116
747C(Dual)	0	+70	20	150 k	85	200	500	±12	0.5	±12	6	yes	yes	yes	yes	TO-101, 116
748B	−55	+125	50	300 k	85	200	500	±12	0.5	±12	5	yes	no	yes	yes	TO-99
748C	0	+70	20	150 k	85	200	500	±12	0.5	±12	6	yes	no	yes	yes	TO-99
749B(Dual)	−55	+125	25	100 k	220	400	750	±11	1.5	±12, −14.5	3	no	no	yes	yes	TO-116
749C(Dual)	0	+70	15	70 k	330	500	1000	±11	1.5	±12, −14.5	6	no	no	yes	yes	TO-116
800B	−55	+125	10	250 k	180	100	1000	±4	—	±6	50	no	no	no	no	TO-101
800D	−55	+125	10	100 k	180	200	2000	±4	—	±12	—	no	no	no	no	TO-101
801B	−55	+125	10	250 k	180	100	1000	±4	—	±12	50	no	no	no	no	TO-100
801D	−55	+125	10	100 k	180	200	2000	±4	—	±12	—	no	no	no	no	TO-100
805B	−55	+125	30	500 k	225	50	500	±8	2.5	±12	5	no	no	no	yes	TO-91, 99
805C	0	+100	10	100	225	100	1000	±8	2.5	±12	10	no	no	no	yes	TO-91, 99
806B	−55	+125	30	500	225	50	500	±8	2.5	±9	5	no	no	no	yes	TO-91, 99
806C	0	+100	10	100	225	100	1000	±8	2.5	±9	10	no	no	no	yes	TO-91, 99
807B	−55	+125	30	500	225	50	500	±8	2.5	±12	2.5	no	no	no	yes	TO-91, 99
808A	−55	+125	25	1 M	225	15	50	±8	2.5	±12	5	no	no	no	yes	TO-91, 99
808B	−55	+125	25	1 M	225	30	50	±8	2.5	±12	10	no	no	no	yes	TO-91, 99
809B	0	+125	10	100	150	100	500	±10	—	±10	10	no	no	yes	yes	TO-99, 116
809C	0	+100	10	50	150	350	1000	±10	—	±10	10	no	no	yes	yes	TO-99, 116
810B(Dual)	−55	+125	10	100	150	100	500	±10	—	±10	10	no	no	yes	yes	TO-116
810C(Dual)	0	+100	10	50	150	350	1000	±10	—	±10	10	no	no	yes	yes	TO-116

TABLE E-1. (Continued)

Device	Operating Temp. Range °C Min	Max	A_{VOL} Min K	R_i Min Ω	P_D mW	$|I_{io}|$ Max nA	I_b Max na	CMV_i Min V	*Typ. Slew Rate SR V/μs	V_o Min V	$|V_{io}|$ Max mV	Offset Adjust	Internal Compensation	Output Protection	Input Protection	JEDEC Package Type
715B	−55	+125	150	1M(typ)	210	250	750	±15	18	±10	±5	yes	yes	yes	yes	TO-100
715C	0	+70	10	1M(typ)	300	1500	250	±15	18	±10	±7.5	yes	no	yes	yes	TO-100
846B	−55	+125	100	25 M	90	5	30	±12.5	2.0	±12	±3	yes	no	yes	yes	TO-99
846C	0	+70	50	15 M	75	15	50	±12	2.0	±12	±5	yes	no	yes	yes	TO-99
LM101A	−55	+125	50	1.5 M	120	10	25	±12	0.5	±12	±2	yes	no	yes	yes	TO-99
LM101B	−55	+125	50	300 k	120	200	500	±12	0.5	±12	±5	yes	no	yes	yes	TO-99
LM201A	−25	+85	50	1.5 M	120	10	75	±12	0.5	±12	±2	yes	no	yes	yes	TO-99
LM201C	−25	+85	20	300 k	90	200	500	±12	0.5	±12	±7.5	yes	no	yes	yes	TO-99
LM301A	0	+70	25	500 k	120	50	250	±12	0.5	±12	±7.5	yes	no	no	yes	TO-99
LM307D	0	+70	25	500 k	120	50	250	±12	0.5	±12	±7.5	yes	no	no	yes	TO-99
811B	−55	+125	10	100	150	100	500	±10	—	±10	10	no	no	yes	yes	TO-99, 116
811C	0	+100	10	50	—	350	1000	±10	—	±10	10	no	no	yes	yes	TO-99, 116
813C	0	+70	6	—	120	2000	5000	±5	—	±4	4	no	no	yes	yes	TO-99, 116
819B	−55	+125	5	50 k	25	100	500	±4	0.5	±12	10	no	no	yes	yes	TO-99
841B	−55	+125	50	300	85	200	500	±12	0.5	±12	5	no	no	yes	yes	TO-99
841C	0	+100	20	150 k	85	200	500	±12	0.5	±12	6	no	no	yes	yes	TO-99
844B	−55	+125	100	25 M	75	5	30	±12.5	2.0	±12	±3	yes	yes	yes	yes	TO-99
844C	0	+70	50	15 M	90	15	50	±12	2.0	±12	±5	yes	yes	yes	yes	TO-99
LM107B	−55	+125	50	1.5 M	120	10	75	±12	0.5	±12	±2	yes	yes	yes	yes	TO-99
LM207C	−25	+85	50	1.5 M	120	10	75	±12	0.5	±12	±2	yes	yes	yes	yes	TO-99
MC1437C(Dual)	0	+70	15	50 k	200	500	1500	±8	0.3	±10	10	no	no	no	no	TO-116
MC1439C	0	+70	15	100	200	100	1000	±11	4.2	±10	7.5	no	no	yes	yes	TO-99, 116
MC1458C(Dual)	0	+70	20	150 k	85	200	500	±12	0.5	±12	6	no	yes	yes	yes	TO-99, 116
MC1537B(Dual)	−55	+125	25	150 k	165	200	500	±8	0.3	±12	5	no	no	no	no	TO-91, 99, 116
MC1539B	−55	+125	50	150	150	60	500	±11	4.2	±10	3	no	no	yes	yes	TO-99, 116
MC1558B(Dual)	−55	+125	50	300 k	85	200	500	±12	0.5	±12	5	yes	yes	yes	yes	TO-101, 116

* Unity gain.

TABLE E-2

Commercial Temperature Range: $0°C \leq T_A \leq +70°C$

Device	Input Offset Voltage Max (mV)	Input Offset Voltage Drift Max ($\mu V/°C$)	Input Offset Current Max (nA)	Input Bias Current Max (nA)	Voltage Gain Min (Volts/V)	Bandwidth $A_V = 1$ Typ (MHz)	Slew Rate $A_V = 1$ Typ ($V/\mu s$)	Output Voltage Swing $R_L = 10\,k\Omega$ (V)	Supply Voltage Min (V)	Supply Voltage Max (V)	Mod Rejection Ratio# (dB) Min	Differential Input Voltage (V)	Supply Current $T_A = 25°C$ Max (mA)	Compensation Components	Package Types
SINGLE OP AMPS															
LM201	10	10 typ	750	200	15 k	1	0.5	5	±3	±22	±12	±30	3	1	TO-5 F.P.
LM301A	10	30	70	300	15 k	1	0.5	5	±3	±18	±12	±30	3	1	TO-5 DIP
LM302	20	20 typ	*	30	0.9985	10	10	1	±12	±18	±10	*	5.5	0	TO-5
LM307	10	30	50	250	15 k	1	0.5	5	±3	±18	±12	±30	3	1	TO-5 DIP F.P.
LM308A	0.73	5	1.5	10	60 k	1	0.3	1	±2	±20	±14	(Note 1)	0.8	1	TO-5 DIP F.P.
LM308	10	30	1.5	10	15 k	1	0.3	1	±2	±18	±14	(Note 1)	0.8	0	TO-5 DIP F.P.
LM310	10	10 typ	*	10	0.999	20	30	1	±5	±18	±10	*	5.5	0	TO-5 DIP F.P.
LM312	10	30	1.5	10	15 k	1	0.3	1	±2	±18	±14	(Note 1)	0.8	0	TO-5 DIP F.P.
LM316A	6	*	0.03	0.1	30 k	1	0.3	1	±5	±18	±13	(Note 1)	0.6	0	TO-5 DIP F.P.
LM316	15	*	0.1	0.25	15 k	1	0.3	1	±5	±20	±13	(Note 1)	0.8	0	TO-5 DIP
LM318	15	*	300	750	20 k	15	50	5	±5	±18	±11.5	(Note 1)	10	0	TO-5 DIP
LM321A ($R_{SET} = 70k$)	0.65	0.2	1	25	12 k	0.5	*	*	±5	±20	±15	±15	2.2	1	TO-5 DIP F.P.
LM321 ($R_{SET} = 70k$)	2.5	1	4	28	12 k	0.5	*	*	±5	±20	±15	±15	2.2	1	TO-5 DIP F.P.
LM343	10	*	14	55	50 k	1	2.5	4 ($R_L \geq 5\,k$)	±4	±34	±34	±34	5.0	0	TO-5 DIP F.P.
LM344	10	*	14	55	50 k	2	30	4 ($R_L \geq 5\,k$)	±4	±34	±34	±34	5.0	1	TO-5 DIP F.P.
LF351	10	10 typ	0.1	0.2	25 k	4	13	12	−18	18	70	±30	3.4	0	H.N
LF351A	2	10 typ	0.05	0.2	50 k	4	13	±12	−18	18	80	±30	2.8	0	H.N
LF351B	5	10 typ	0.1	0.1	50 k	4	13	±12	−18	18	80	±30	2.8	0	H.N
LF355A	2.3	5	0.1	5	25 k	2.5	5	5	±5	±22	±20	±40	4	0	TO-5, Mini-DIP
LF355	13	5 typ	2	8	15 k	2.5	5	5	±5	±18	±16	±30	4	0	TO-5, Mini-DIP
LF356A	2.3	5	1	5	25 k	5	15	5	±5	±22	±20	±40	10	0	TO-5, Mini-DIP
LF356	13	5 typ	2	8	15 k	5	15	5	±5	±18	±16	±30	10	0	TO-5, Mini-DIP
LF357A ($A_V \geq 5$)	2.3	5	1	5	25 k	25	75	5	±5	±22	±20	±40	10	0	TO-5, Mini-DIP
LF357 ($A_V \geq 5$)	13	5 typ	2	8	15 k	25	75	5	±5	±18	±16	±30	10	0	TO-5, Mini-DIP
LF13741	20	10 typ	2	8	15 k	1	0.5	5	±4	±18	±16	±30	4	0	TO-5, Mini-DIP

Note 1: Inputs have shunt-diode protection; current must be limited.

*Not specified

#This column shows either the maximum common-mode input voltage or the minimum CMRR.

APPENDIX F

SPECIFICATIONS OF THE 741 OP AMP

	Parameters	Minimum	Typical	Maximum
A_{VOL}	Open loop voltage gain	50,000	200,000	
V_{io}	Input offset voltage		1 mV	6 mV
I_B	Input bias current		80 nA	500 nA
I_{io}	Input offset current		20 nA	200 nA
R_o	Output resistance		75 Ω	
CMR (dB)	Common mode rejection	70 dB	90 dB	
R_i	Input resistance	300 kΩ	2 MΩ	

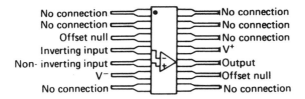

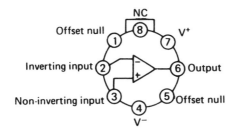

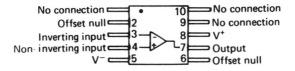

Figure F-1

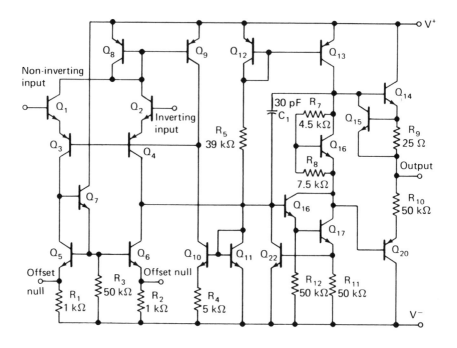

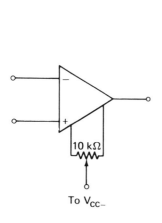

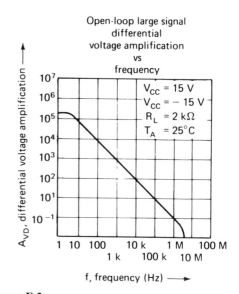

Figure F-2

APPENDIX G

DERIVATION OF EQ. (4-4)

As shown in the circuit of Fig. G-1, the bias current I_{B_1} flows through two paths, R_1 and R_F. This current sees the resistors R_1 and R_F in parallel and causes a voltage drop across them, which is the voltage at the inverting input 1 to ground. This can be shown as

$$V_1 = \left(\frac{R_1 R_F}{R_1 + R_F} \right) I_{B_1}.$$

The current I_{B_2} sees no resistance between the noninverting input 2 and ground, and therefore the voltage at this input is $V_2 = 0$ V to ground. Thus the differential input voltage caused by these currents is $V_1 - V_2 = V_1$. This voltage V_1 is amplified by the circuit's closed-loop gain, resulting in an output

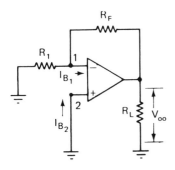

Figure G-1

offset of

$$V_{oo} = A_v V_1 \cong -\frac{R_F}{R_1} V_1 = -\frac{R_F}{R_1} \left(\frac{R_1 R_F}{R_1 + R_F} \right) I_{B_1} = \frac{-R_F R_F}{R_1 + R_F} (I_{B_1}).$$

Since R_F^2 is much larger than $R_1 + R_F$, the above simplifies to

$$V_{oo} \cong \frac{R_F^2}{R_F} (I_{B_1}) = R_F I_{B_1}. \qquad (4\text{-}4)$$

APPENDIX H

DERIVATION OF EQ. (4-7)

As shown in the circuit of Fig. H-1, the bias circuit I_{B_1} flows through two paths, R_1 and R_F. This current sees essentially resistors R_1 and R_F in parallel and causes a voltage drop across them and at the input 1 with respect to ground; that is,

$$V_1 = \left(\frac{R_1 R_F}{R_1 + R_F} \right) I_{B_1}.$$

The current I_{B_2} flows through resistance R_2, causing input 2 to be off ground by

$$V_2 = R_2 I_{B_2} = \left(\frac{R_1 R_F}{R_1 + R_F} \right) I_{B_2}.$$

The difference in these voltages, $V_1 - V_2$, is the differential input which is amplified by the circuit's closed-loop gain—that is,

$$V_{oo} = A_v(V_1 - V_2) = -\frac{R_F}{R_1}\left[\frac{R_1 R_F}{R_1 + R_F}(I_{B_1}) - \frac{R_1 R_F}{R_1 + R_F}(I_{B_2}) \right]$$

$$= \frac{R_F^2}{R_1 + R_F}(I_{B_1} - I_{B_2}).$$

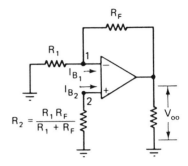

Figure H-1

Since R_F^2 is much larger than $R_1 + R_F$ and since $I_{B_1} - I_{B_2} = I_{io}$, then

$$V_{oo} \cong R_F I_{io}, \qquad (4\text{-}7)$$

where V_{oo} is either positive or negative, depending on whether I_{B_1} is the larger or I_{B_2} is the larger of the two bias currents.

APPENDIX I

DERIVATION OF EQ. (6-1)

Figure 2-4 of Chapter 2 shows that an Op Amp is equivalent to a signal source $(A_{VOL})V_{id}$ in series with a resistance (R_o), as the load (R_L) "sees" it. An Op Amp also has internal capacitances. Their effect on the Op Amp's output characteristics is the same as the effect of a single *equivalent* capacitance C that is across the output and ground (see Fig. I-1). At low frequencies, the reactance of this capacitance C is so large that it acts like an open and has no influence on the amplitude of the output signal voltage V_o. At higher frequencies, however, the reactance of C decreases, causing the amplitude of the output V_o to decrease. A typical V_o vs. frequency curve is shown in Fig. I-2. Such curves can be plotted by measuring V_o with various frequencies of input signals V_s, while the amplitude of the input signal V_s is kept constant. Note that f_c is the frequency at which V_o decreases to 0.707 of its maximum value. This is a 3-dB decrease. Frequency f_c is called the *critical frequency*, *cutoff frequency*, *half-power frequency*, or *3-dB-down frequency*, to name just a few of the more common terms. As shown in Fig. I-2, f_c is the upper limit of an amplifier's bandwidth.

In the equivalent circuit of Fig. I-1, the reactance of C is equal to the resistance of R at the frequency f_c. Since, generally,

$$X_c = \frac{1}{2\pi fC},$$

then

$$X_c = R = \frac{1}{2\pi f_c C},$$

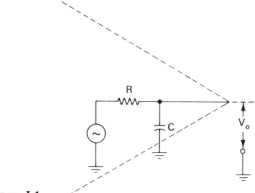

Figure I-1

and therefore,

$$f_c = \frac{1}{2\pi RC}.$$ (9-6)

Since Op Amps can amplify down to 0 Hz and f_c marks the upper limit of the bandwidth, we can show that the bandwidth $BW = f_c$. The above equation can thus be modified to

$$BW = \frac{1}{2\pi RC}.$$ (I-1)

If a relatively high frequency square wave is applied to the input of an amplifier, the equivalent circuit and resulting output are as shown in Fig. I-3a and b. Note that V_o rises to $0.1V_{max}$ at time t_1 and that V_o reaches $0.9V_{max}$ at

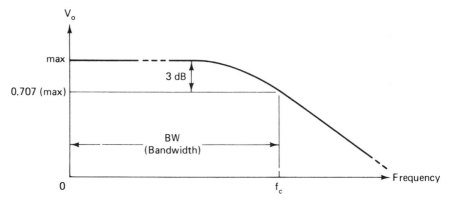

Figure I-2

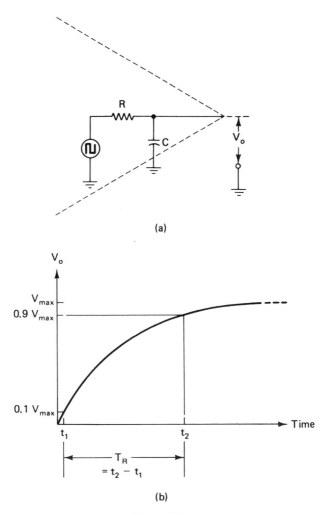

(a)

(b)

Figure I-3

t_2. Therefore, the rise time

$$T_R = t_2 - t_1.$$ (I-2)

Since this output rises exponentially, it can generally be expressed with the equation

$$V_{o(t)} = V_{max}(1 - e^{-t/RC}).$$

Specifically at time t_1, then,

$$0.1V_{max} = V_{max}(1 - e^{-t_1/RC}).$$ (I-3)

Dividing both sides of Eq. (I-3) by V_{max} yields

$$0.9 = e^{-t_1/RC}.$$

Finding the natural logarithm of both sides shows that

$$0.1 \cong t_1/RC,$$

and, therefore,

$$t_1 \cong 0.1RC. \tag{I-4}$$

Similarly, at t_2,

$$0.9V_{max} = V_{max}(1 - e^{-t_2/RC}).$$

Dividing by V_{max} and rearranging yields

$$0.1 = e^{-t_2/RC},$$

and, thus

$$2.3 \cong t_2/RC$$

and

$$t_2 \cong 2.3RC. \tag{I-5}$$

By substituting Eqs. (I-4) and (I-5) into Eq. (I-2) we can show that

$$T_R \cong 2.3RC - 0.1RC = 2.2RC,$$

and thus

$$RC \cong T_R/2.2.$$

This last equation substituted into Eq. (I-1) gives us

$$BW \cong \frac{1}{2\pi T_R/2.2}$$

or

$$BW \cong \frac{0.35}{T_R}. \tag{6-1}$$

Similar algebraic steps will show that also $BW \cong 0.35/T_F$.

APPENDIX J

SPECIFICATIONS OF THE
709 OP AMP

	Parameter	Minimum	Typical	Maximum
A_{VOL}	Open loop voltage gain	15,000	45,000	
V_{io}	Input offset voltage		1 mV	7.5 mV
I_B	Input bias current		200 nA	1500 nA
I_{io}	Input offset current		50 nA	500 nA
R_o	Output resistance		150 Ω	
CMR (dB)	Common mode rejection	65 dB	90 dB	
R_i	Input resistance	50 kΩ	250 kΩ	

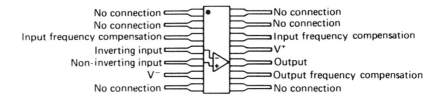

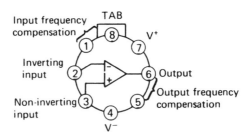

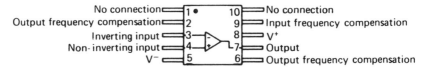

Figure J-1

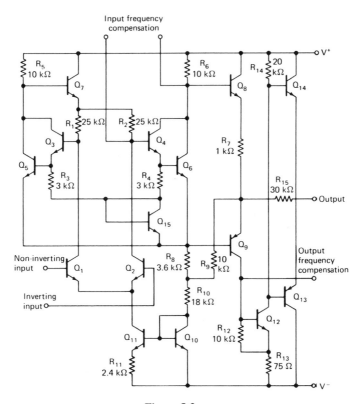

Figure J-2

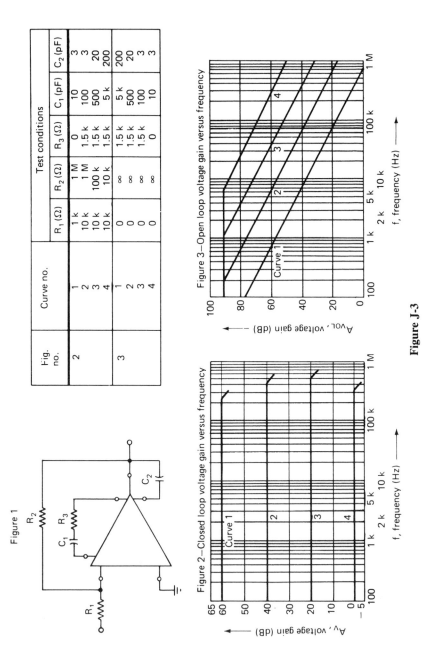

Figure 1

Figure 2—Closed loop voltage gain versus frequency

Figure 3—Open loop voltage gain versus frequency

Fig. no.	Curve no.	Test conditions					
		$R_1 (\Omega)$	$R_2 (\Omega)$	$R_3 (\Omega)$	$C_1 (pF)$	$C_2 (pF)$
2	1	1 k	1 M	0	10	3
	2	10 k	1 M	1.5 k	100	3
	3	10 k	100 k	1.5 k	500	20
	4	10 k	10 k	1.5 k	5 k	200
3	1	0	∞	1.5 k	5 k	200
	2	0	∞	1.5 k	500	20
	3	0	∞	1.5 k	100	3
	4	0	∞	0	10	3

Figure J-3

APPENDIX K

SELECTION OF COUPLING CAPACITORS

The value of coupling capacitance C used between ac amplifier stages is determined by the required low-frequency response and the dynamic output and input resistances of the stages being coupled. Generally, lower frequencies and smaller resistances require larger coupling capacitors.

If f_1 is the required low end of the bandwidth (that is, the low frequency at which the signal being coupled is 0.707 of its value at medium frequencies), then

$$C = \frac{1}{2\pi f_1(R_o + R_i)},$$

where C is the capacitance between the two stages being coupled

R_o is the effective ac output resistance of the first stage, and

R_i is the effective ac input resistance of the second stage.

APPENDIX L

OP AMP AS AN INTEGRATOR

The term operational amplifier originated when Op Amps were developed to perform mathematical calculations in analog computers. Although analog computers have been displaced by digital computers, the principles of calculations in analog systems are still useful and used. In Section 8.3 we observed how an integrator responds to square and rectangular waveforms. These are common and simple examples of integrator applications. Other input waveforms can be integrated too. Referring back to Eq. (8-10d),

$$\Delta V_o = \frac{I}{C} \Delta t.$$

We note that C is a constant and then can integrate both sides to get

$$V_o = -\frac{1}{C} \int I \, dt.$$

Recall that I is the current being forced through the capacitor. Substituting the right side of Eq. (8-9) into the above results in

$$V_o = -\frac{1}{C} \int \frac{V_s}{R} \, dt$$

or

$$V_o = -\frac{1}{RC} \int V_s \, dt,$$

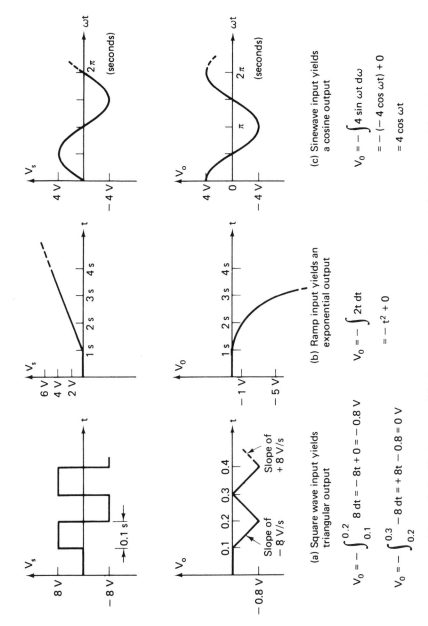

(a) Square wave input yields triangular output

$$V_0 = -\int_{0.1}^{0.2} 8\, dt = -8t + 0 = -0.8\ V$$

$$V_0 = -\int_{0.2}^{0.3} -8\, dt = +8t - 0.8 = 0\ V$$

(b) Ramp input yields an exponential output

$$V_0 = -\int 2t\, dt$$

$$= -t^2 + 0$$

(c) Sinewave input yields a cosine output

$$V_0 = -\int 4\sin \omega t\, d\omega$$

$$= -(-4\cos \omega t) + 0$$

$$= 4\cos \omega t$$

Figure L-1 Various possible input and output waveforms of the circuit in Fig. 8-9 where $R \times C = 1$.

303

where the negative sign represents the phase-inverting property of this circuit. If we select the R and C values so that their product is 1, the previous equation simplifies to

$$V_o = -\int V_s \, dt.$$

See Fig. L-1 for some typical V_o vs. V_s waveforms.

APPENDIX M

OP AMP AS A DIFFERENTIATOR

The Op Amp can be wired to work as a differentiator. As such, its output is essentially the derivative of the input voltage waveform. The basic differentiator is the circuit shown in Fig. M-1a. The right side of the input capacitor C is virtually grounded, and therefore the voltage across it is the input voltage V_s. Current I flows to charge or discharge C only when the input voltage changes. Thus, by substituting V_s for v_c and I for i_c in Eq. (8-10a), we can show that

$$I = C \frac{dV_s}{dt}. \tag{M-1}$$

Since the resistance looking into the inverting input is very large, any existing input current I is forced up through the feedback resistor R_F. But the left side of R_F is virtually grounded, and therefore the voltage drop across it is the output voltage V_o. By Ohm's law, then,

$$V_o = R_F I.$$

Rearranging and substituting this last equation into Eq. (M-1), we can show that

$$\frac{V_o}{R_F} = -C \frac{dV_s}{dt}$$

or

$$V_o = -R_F C \frac{d}{dt}(V_s). \tag{M-2}$$

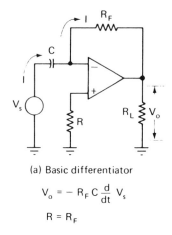

(a) Basic differentiator

$$V_o = - R_F C \frac{d}{dt} V_s$$

$$R = R_F$$

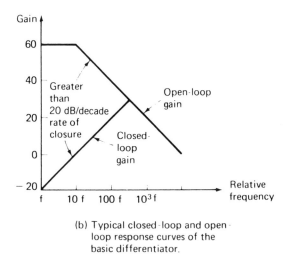

(b) Typical closed-loop and open-
loop response curves of the
basic differentiator.

Figure M-1

Apparently, the output voltage V_o is the derivative of the input voltage V_s times the negative product of R_F and C.

Unfortunately, the circuit in Fig. M-1 has some practical problems. Because the ratio of the feedback resistance R_F to the input capacitor's reactance, X_C rises with higher frequencies, this circuit's gain increases with frequency. This tends to amplify the high-frequency noise generated in the system, and the resulting output noise can completely override the differentiated signal.

Another problem with this basic differentiator is its tendency to be unstable. In Chapter 6 we learned that, if the closed-loop and open-loop gain

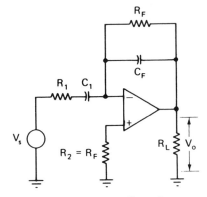

(a) Practical differentiator

$$f_x = \frac{1}{2\pi R_F C_1}$$

$$f_y = \frac{1}{2\pi R_1 C_F}$$

f_z is the frequency at which the
open loop-gain is unity

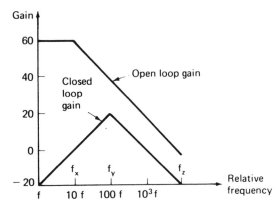

(b) Typical closed-loop and open-
loop response curves of the
practical differentiator.

Figure M-2

vs. frequency curves intersect at a rate of closure greater than 20 dB/decade, the circuit may be unstable. In the case of the circuit in Fig. M-1a, the input capacitor C causes a low-frequency roll-off that intersects the open-loop curve at a rate of closure greater than 20 dB/decade as shown in Fig. M-1b. This means that the circuit will probably be unstable. The problem can be solved if we add two components as shown in Fig. M-2a. Components R_1 and C_1 cause a 20-dB/decade roll-off with decreasing frequencies, while R_F and C_F cause a -20-dB/decade roll-off at higher frequencies, resulting in a

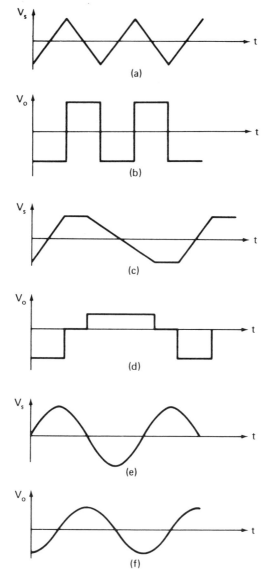

Figure M-3 Differentiator's input and resulting output waveforms; input waveform (a) causes output waveform (b); input waveform (c) causes output waveform (d); and input waveform (e) causes output waveform (f).

closed-loop gain vs. frequency curve as shown in Fig. M-2b. The components R_1, R_F, and C_F can be selected so that the closed-loop gain vs. frequency curve does *not* intersect with the open-loop curve, thus assuring us of stable operation.

Some input voltages V_s and the resulting differentiated outputs V_o are shown in Fig. M-3.

APPENDIX N

PHASE SHIFT IN SECOND-ORDER ACTIVE FILTERS

The amplitude of the output V_{out} vs. frequency is usually most important in active filter applications. Where phase shift ϕ between V_{in} and V_{out} is also of concern, the phase angle of V_{out} with respect to V_{in} is

$$\phi = -\arctan\left[\frac{2\pi f(R_1C_2 + R_2C_2 + R_1C_1(1 - A_v))}{\left|1 - 2\pi f\sqrt{R_1R_2C_1C_2}\right|}\right],$$

with the LP filter types of Figs. 9-9 and 9-15, and where

$$A_v = \frac{R_f}{R_i} + 1.$$

With HP filter types of Figs. 9-12 and 9-16, the phase shift

$$\phi = -\arctan\left[\frac{2\pi f(R_1C_1 + R_1C_2 + R_2C_2(1 - A_v))}{\left|1 - 2\pi f\sqrt{R_1R_2C_1C_2}\right|}\right].$$

The BASIC programs of Appendix O include solutions of phase shift ϕ.

APPENDIX O

PROGRAMS THAT DISPLAY TABLES OF GAIN vs. FREQUENCY OF SECOND-ORDER LOW-PASS AND HIGH-PASS ACTIVE FILTERS

```
10 PRINT"*****************************************"
20 PRINT"*   THIS PROGRAM WILL PRINT A TABLE    *"
30 PRINT"*   OF GAIN VS FREQUENCY FOR A 2ND     *"
40 PRINT"*    ORDER LOW PASS ACTIVE FILTER      *"
50 PRINT"*****************************************"
60 PRINT:PRINT
70 PRINT"WHAT IS THE VALUE OF R1 IN OHMS":INPUT R1
80 PRINT"WHAT IS THE VALUE OF R2 IN OHMS":INPUT R2
90 PRINT"WHAT IS THE VALUE OF C1 IN FARADS":INPUT C1
100 PRINT"WHAT IS THE VALUE OF C2 IN FARADS":INPUT C2
110 PRINT"WHAT IS THE CLOSED-LOOP GAIN":INPUT B
120 PRINT
130 REM    *****************************************
140 REM    *   SOLVING FOR CUTOFF FREQ AND THE    *
150 REM    *        DAMPING COEFFICIENT           *
160 REM    *****************************************
170 F1=1/(2*3.141593*(R1*R2*C1*C2)^.5)
180 D=1/(R1*R2*C1*C2)^.5
190 A=(R2*C2+R1*C2+R1*C1*(1-B))*D
200 REM    *****************************************
210 REM    *   CALCULATING CUTOFF FREQ AND THE    *
220 REM    *        DAMPING COEFFICIENT           *
230 REM    *****************************************
240 PRINT"YOUR CUTOFF FREQUENCY IS";F1;"HZ"
250 PRINT"YOUR DAMPING COEFFICIENT D = ";A
260 PRINT"DO YOU WANT TO PROCEED. ANSWER Y OR N":INPUT A$
270 IF A$="Y" OR A$="y" GOTO 350
280 PRINT"DO YOU WANT TO ENTER NEW R & C VALUES. (Y/N)":INPUT A$
290 IF A$="Y" OR A$="y" GOTO 70
300 GOTO 670
310 REM    *****************************************
320 REM    *   OBTAINING RANGE AND INCREMENTS     *
330 REM    *          OF TABLE LISTING            *
340 REM    *****************************************
350 PRINT"WHAT IS YOUR INITIAL (LOWEST) FREQUENCY":INPUT F
360 PRINT"WHAT IS YOUR FINAL (HIGHEST) FREQUENCY":INPUT T
370 PRINT"BY WHAT MULTIPLES DO YOU WHAT THIS FREQUENCY"
380 PRINT"TO INCREASE":INPUT M
390 PRINT
400 REM    *****************************************
410 REM    *    PRINT THE HEADER OF THE TABLE     *
420 REM    *****************************************
430 PRINT"FREQ";TAB(15);"GAIN";TAB(25);"GAIN";TAB(35);"PHASE"
440 PRINT"  HZ";TAB(14);"OUT/IN";TAB(26);"DB";TAB(35)"DEGREES"
450 PRINT"..............................................."
```

Figure O-1

```
460 REM    ****************************************
470 REM    * GAIN AND PHASE SHIFT CALCULATIONS  *
480 REM    ****************************************
490 W=2*3.141593*F*(R1*R2*C1*C2)^.5
500 K=ABS(1-W^2)
510 G=B/(K^2+(A*W)^2)^.5
520 G1=20/2.302585*LOG(G)
530 O=-ATN(A*W/K)*57.29578
540 F3=INT(F*100!)/100!:G=INT(G*10000!)/10000!
550 G1=INT(G1*100!)/100!:O=INT(O*1000!)/1000!
560 REM    ****************************************
570 REM    * LISTING GAIN & PHASE VS FREQUENCY   *
580 REM    ****************************************
590 PRINT F3:TAB(15);G;TAB(25);G1;TAB(35);O
600 F=F*M
610 IF F<(T*M) THEN GOTO 490
620 PRINT"DO YOU WANT TO RUN WITH THIS FILTER AGAIN. (Y/N)":INPUT A$
630 IF A$="Y" OR A$="v" GOTO 350
640 PRINT"DO YOU WANT TO ENTER NEW COMPONENT VALUES. (Y/N)":INPUT A$
650 IF A$="Y" OR A$="v" GOTO 70
660 PRINT"THANK YOU AND GOODBY!"
670 END
O
```

Figure O-1 *(Continued)*

```
10 PRINT"****************************************"
20 PRINT"*  THIS PROGRAM WILL PRINT A TABLE    *"
30 PRINT"*   OF GAIN VS FREQUENCY FOR A 2ND    *"
40 PRINT"*   ORDER HIGH PASS ACTIVE FILTER     *"
50 PRINT"****************************************"
60 PRINT:PRINT
70 PRINT"WHAT IS THE VALUE OF R1 IN OHMS":INPUT R1
80 PRINT"WHAT IS THE VALUE OF R2 IN OHMS":INPUT R2
90 PRINT"WHAT IS THE VALUE OF C1 IN FARADS":INPUT C1
100 PRINT"WHAT IS THE VALUE OF C2 IN FARADS":INPUT C2
110 PRINT"WHAT IS THE CLOSED-LOOP GAIN":INPUT B
120 PRINT
130 F1=1/(2*3.141593*(R1*R2*C1*C2)^.5)
140 D=1/(R1*R2*C1*C2)^.5
150 A=(R1*C1+R1*C2+R2*C2*(1-B))*D
160 PRINT"YOUR CUTOFF FREQUENCY IS";F1;"HZ"
170 PRINT"YOUR DAMPING COEFFICIENT D = ";A
180 PRINT"DO YOU WANT TO PROCEED. ANSWER Y OR N":INPUT A$
190 IF A$="Y" OR A$="v" GOTO 230
200 PRINT"DO YOU WANT TO ENTER NEW R & C VALUES. Y OR N":INPUT A$
210 IF A$="Y" OR A$="v" GOTO 70
220 GOTO 460
230 PRINT"WHAT IS YOUR INITIAL (HIGHEST) FREQUENCY":INPUT F
240 PRINT"WHAT IS YOUR FINAL (LOWEST) FREQUENCY":INPUT T
250 PRINT"BY WHAT MULTIPLES DO YOU WHAT THIS FREQUENCY"
260 PRINT"TO DECREASE":INPUT M
270 PRINT
280 PRINT"FREQ":TAB(15):"GAIN":TAB(25):"GAIN":TAB(35):"PHASE"
290 PRINT"  HZ":TAB(14):"OUT/IN":TAB(26):"DB":TAB(35)"DEGREES"
300 PRINT"........................................."
310 W=2*3.141593*F*(R1*R2*C1*C2)^.5
320 K=ABS(1-W^2)
330 G=B*W^2/(K^2+(A*W)^2)^.5
340 G1=20/2.302585*LOG(G)
350 O=-ATN(A*W/K)*57.29578
360 F3=INT(F*100!)/100!:G=INT(G*10000!)/10000!
370 G1=INT(G1*100!)/100!:O=INT(O*1000!)/1000!
380 PRINT F3:TAB(15):G;TAB(25):G1;TAB(35):O
390 F=F/M
400 IF F>(T/M) THEN GOTO 310
410 PRINT"DO YOU WANT TO RUN WITH THIS FILTER AGAIN (Y/N)":INPUT A$
420 IF A$="Y" OR A$="v" GOTO 230
430 PRINT"DO YOU WANT TO ENTER NEW COMPONENT VALUES (Y/N)":INPUT A$
440 IF A$="Y" OR A$="v" GOTO 70
450 PRINT"THANK YOU AND GOODBY!"
460 END
O
```

Figure O-2

APPENDIX P

PROGRAM FOR ANALYZING MULTIPLE-FEEDBACK BANDPASS FILTERS

```
10 ' ********************
20 ' Gain vs. Frequency Program: This program displays a table of Gain vs.
30 ' Frequency characteristics of a multiple-feedback bandpass active filter.
40 ' ********************
50 ' ----------
60 ' Input Data
70 ' ----------
80 KEY OFF:WIDTH 80:SCREEN 0,0,0:CLS
90 PRINT TAB(9);"This program will display a table of gain vs. frequency using"
100 PRINT TAB(7);"the characteristics of a multiple-feedback bandpass active fil
ter."
110 PRINT "------------------------------------------------------------------
-----------"
120 PRINT "Please enter your component values in below:"
130 PRINT
140 INPUT "Enter the value of R1 in ohms ";R1
150 INPUT "Enter the value of R2 in ohms ";R2
160 INPUT "Enter the value of the Feedback Resistor ";RF
170 INPUT "If all the capacitor are equal, enter the value of C ";C
180 PRINT
190 PRINT "The table will be displayed using the range of frequency you enter be
low:
200 FC=1/(2*3.14159)*SQR((R1+R2)/(R1*R2*RF*C^2))
210 A$="Min.F< "+STR$(INT(FC*10+.5)/10)+" <Max.F"
220 PRINT TAB((80-LEN(A$))/2);A$
230 PRINT
240 INPUT "Enter the minimum (or initial) frequency in hertz   ";FSTART
250 INPUT "Enter the maximum (or ending ) frequency in hertz   ";FEND
260 INPUT "Enter the increment to step the frequency in hertz ";FINC
270 ' --------------------
280 ' Calculate Fixed Data
290 ' --------------------
300 O=2*3.14159
310 BW=O*C*RF
320 Q=FC/(2/BW)
330 D=1/Q
340 A=-RF/(2*R1)
350 S=SQR(1/(4*Q^2)+1)
360 S1=1/(2*Q)
370 F1=FC*(S-S1)
380 F2=FC*(S+S1)
390 W=SQR((R1+R2)/(R1*R2*RF*C^2))
400 LOCATE 24,27:PRINT "Press any key to continue.";
410 A$=INKEY$:IF A$="" THEN 410
420 ' --------------------------------------------------------------------
430 ' Display Data in Table Format & Calculate Variable Data
440 ' --------------------------------------------------------------------
450 CLS
460 PRINT "The center frequency is ";FC;"Hz"
470 PRINT "The bandwidth is ";BW;"Hz"
480 PRINT "The Q is ";Q
490 PRINT "F1=";F1;"Hz and F2=";F2;"Hz"
500 PRINT
510 PRINT " Frequency              Gain                 Gain(dB)"
520 PRINT "-----------------------------------------------------------"
530 FOR F=FSTART TO FEND STEP FINC
540 W1=O*F
550 GAIN=SQR((A*D*W*W1)^2/(W1^4+W^2*(D^2-2)*W1^2+W^4))
560 GAINDB=20/2.30259*LOG(GAIN)
570 A$="  #########.#         ######.##          ######.##"
580 IF GAIN>999999.99# OR GAINDB>999999.99# THEN A$="  #########.#         #.###^^
^^         #.###^^^^"
590 PRINT USING A$;F,GAIN,GAINDB
600 NEXT F
610 END
```

Figure P-1

APPENDIX Q

PROGRAM FOR ANALYZING MULITPLE-FEEDBACK BANDSTOP FILTERS

Program in BASIC that prints or displays a table of gain vs. frequency characteristics of a multiple-feedback bandsop (notch) filter.

```
list
10 ' *******************
20 ' Gain vs. Frequency Program: This program displays a table of Gain vs.
30 ' Frequency characteristics of a multiple-feedback bandstop filter.
40 ' *******************
50 ' 09/14/87  Modified -> 12/03/87
60 DEFDBL G,W,C
70 SCREEN 0,0,0:CLS
80 COLOR 7,3
90 LOCATE ,28:PRINT "Bandstop Filter Program"
100 COLOR 7,0
110 LOCATE 5,20:PRINT "ZDDDDDDDDDDDDDDDDDDDDDDDDDDDDDDDDDDDDDDDDDD?"
120 LOCATE ,20 :PRINT "3                                          3"
130 LOCATE ,20 :PRINT "3      1 - Compute the Filter             3"
140 LOCATE ,20 :PRINT "3      2 - Change current frequency        3"
150 LOCATE ,20 :PRINT "3      3 - Graph the Gain (dB) vs.        3"
160 LOCATE ,20 :PRINT "3      4 - Display Data                   3"
170 LOCATE ,20 :PRINT "3      5 - Exit                           3"
180 LOCATE ,20 :PRINT "3                                          3"
190 LOCATE ,20 :PRINT "3                                          3"
200 LOCATE ,20 :PRINT "3                                          3"
210 LOCATE ,20 :PRINT "@DDDDDDDDDDDDDDDDDDDDDDDDDDDDDDDDDDDDDDDDDDY"
220 LOCATE 14,28:INPUT "Enter (1-6): ",A
230 IF A<1 OR A>6 THEN BEEP:LOCATE 14,28:PRINT SPACE$(20);:GOTO 220
240 ON A GOSUB 330,540,1010,1270,1350
250 LOCATE 24,26:PRINT "( Press any key to continue )";
260 WHILE INKEY$=""
270 WEND
280 GOTO 50
290 '
300 ' ----------
310 ' Input Data
320 ' ----------
330 CLS
340 PRINT "This program will compute data for frequency vs. gain characteristics
"
350 PRINT "of a multiple-feedback bandstop filter"
360 PRINT "--------------------------------------------------------------------
----------"
370 LOCATE 6,1:PRINT "Please enter the component value for this filter:"
380 PRINT
390 PRINT "What is the value of R1 in ohms";:INPUT R1
400 PRINT "What is the value of R2 in ohms";:INPUT R2
410 PRINT "What is the value of the feedback resistor RF";:INPUT RF
420 PRINT "What is the value of R5";:INPUT R5
430 PRINT "What is the value of R4";:INPUT R4
440 PRINT "What is the value of R3";:INPUT R3
450 PRINT "What the capacitors equal. What is the value of each capacitor";
460 INPUT C
470 PRINT "What is the minimun (starting) frequency";:INPUT FSTART
480 PRINT "What is the maximum (ending) frequency";:INPUT FEND
490 PRINT "What is the increment frequency";:INPUT FINCREMENT
500 GOTO 690
510 ' -------------------------
520 ' Change the Current Frequency
530 ' -------------------------
540 CLS
550 COLOR 7,3
560 LOCATE ,26: PRINT "Change the Current Frequency"
570 COLOR 7,0
580 LOCATE 5,1:PRINT "Current Starting Frequency =";FSTART
590 LOCATE ,11:PRINT "Ending Frequency =";FEND
600 LOCATE ,8 :PRINT "Increment Frequency =";FINCREMENT
610 PRINT
620 INPUT "Enter the New Starting Frequency: ",FSTART
630 INPUT "                Ending Frequency: ",FEND
640 INPUT "             Increment Frequency: ",FINCREMENT
650 PRINT
```

Figure Q-1

```
660 ' --------------------
670 ' Calculate Fixed Data
680 ' --------------------
690 IF N THEN ERASE F,G,GDB
700 N=(FEND-FSTART)/FINCREMENT+1
710 DIM F(N),G(N),GDB(N)
720 W=SQR((R1+R2)/(R1*R2*RF*C^2))
730 W1=W/(2*3.14159)
740 P=2/(RF*C*2*3.14159)
750 Q=W1/P
760 S=SQR(1/(4*Q^2)+1)
770 S1=1/(2*Q)
780 A=-RF/(2*R1)
790 D=1/Q
800 ' -------------------
810 '     Variable Data
820 ' -------------------
830 I=0
840 PRINT : PRINT "Computing Data .";
850 FOR F=FSTART TO FEND STEP FINCREMENT
860 PRINT ".";
870 W1=2*3.14159*F
880 K=(A*D*W*W1)^2
890 K1=W1^4+W^2*(D^2-2)*W1^2+W^4
900 G=SQR(K/K1)
910 G1=R5/R3-G*R5/R4
920 G1=ABS(G1)
930 G2=20/2.30259*LOG(G1)
940 I=I+1
950 F(I) = F : G(I) = G1 : GDB(I) = G2
960 NEXT F
970 RETURN
980 ' --------------------
990 ' Graph the Notch Filter
1000 ' --------------------
1010 SCREEN 2
1020 LOCATE ,31:PRINT "Notch Filter Graph"
1030 VIEW (173,50) - (453,150),0,1:' Place the Graph in the Middle of the screen
1040 PRINT
1050 PRINT "R1 =";R1;TAB(20);"R2 =";R2;TAB(40);"R3 =";R3;TAB(60);"R4 =";R4
1060 PRINT "R5 =";R5;TAB(20);"RF =";RF;TAB(40);"C  =";USING "###.#^^^^";C
1070 FOR I=1 TO 4:A$=MID$("Gain",I,1):LOCATE 9+I,18:PRINT A$:NEXT I
1080 LOCATE 21,35:PRINT "Frequency"
1090 HIGH=0:LOW=9999
1100 FOR I=1 TO N
1110 IF GDB(I)>HIGH THEN HIGH=GDB(I):HIGH%=I
1120 IF GDB(I)<LOW THEN LOW =GDB(I):LOW% =I
1130 NEXT I
1140 WINDOW (FSTART,HIGH) - (FEND,LOW): ' Set the Scale for the window
1150 PSET (FSTART,HIGH):                ' Set the Starting point for line draw
1160 FOR I=1 TO N
1170 LINE - (F(I),GDB(I))
1180 NEXT I
1190 WHILE INKEY$=""
1200 WEND
1210 ' The above delay is used for Print Screening to the printer with the help
1220 ' of "GRAPHICS" file excuted from DOS before loading this program.
1230 RETURN
1240 ' ----------------------
1250 ' Display the Current Data
1260 ' ----------------------
1270 CLS:PRINT " Frequency           Gain           Gain(dB)"
1280 PRINT "DDDDDDDDDDDDDDDDDDDDDDDDDDDDDDDDDDDDDDDDDDDDDDDDDDDDD"
1290 T$=" ########.#       ######.##         ######.##"
1300 IF G1>999999.99# OR G2>999999.99# THEN T$=" ########.#       #.####^^^^
     #.####^^^^"
1310 FOR I=1 TO N
1320 PRINT USING T$;F(I),G(I),GDB(I)
1330 NEXT I
1340 RETURN
1350 CLS:END
Ok
```

Figure Q-1   *(Continued)*

APPENDIX R

FIXED VOLTAGE REGULATORS

TABLE R-1

Product Type No.	Input Voltage (V) Min	Input Voltage (V) Max	Output Voltage (V) Typ	Load Regulation (mV) Typ	Line Regulation (mV) Typ	Ripple Rejection (dB) Typ	Long Term Stability (mV) Max	Output Noise Voltage (μV) Typ	Quiescent Current (mA)	Operating Temperature Range (°C) Min	Operating Temperature Range (°C) Max	Output Current (A)	Package Type
Positive Regulators													
LM109K	7	35	5	50	4	75	10	40	6	-55	125	>1	TO-3, TO-5
LM209K	7	35	5	50	4	75	10	40	6	-25	85	>1	TO-3, TO-5
LM309K	7	35	5	50	4	75	20	40	6	0	70	>1	TO-3, TO-5
LM123K	7.5	20	5	50	4	75	10	40	6	-55	125	3	TO-3
LM223K	7.5	20	5	50	4	75	10	40	6	-25	85	3	TO-3
LM323K	7.5	20	5	50	4	75	20	40	6	0	70	3	TO-3
LM340-05	7	35	5	100 max	100 max	70	20	40	6	0	70	>1	TO-3, TO-220
LM340-06	8	35	6	120 max	120 max	65	24	45	6	0	70	>1	TO-3, TO-220
LM340-08	10	35	8	160 max	160 max	62	32	52	6	0	70	>1	TO-3, TO-220
LM340-12	14	35	12	240 max	240 max	61	48	75	6	0	70	>1	TO-3, TO-220
LM340-15	17	35	15	300 max	300 max	60	60	90	6	0	70	>1	TO-3, TO-220
LM340-18	20	35	18	360 max	360 max	59	72	110	6	0	70	>1	TO-3, TO-220
LM340-24	26	40	24	480 max	480 max	56	96	170	6	0	70	.8	TO-3, TO-220
Negative Regulators													
LM120K-5	-6	-25	-5	50	10	67	50	150	2	-55	125	>1	TO-3, TO-5
LM220K-5	-6	-25	-5	50	10	67	50	150	2	-25	85	>1	TO-3, TO-5
LM320K-5	-6	-25	-5	50	10	67	50	150	2	0	70	>1	TO-3, TO-5
LM120K-5.2	-6.2	-25	-5.2	50	10	67	50	150	2	-55	125	>1	TO-3, TO-5
LM220K-5.2	-6.2	-25	-5.2	50	10	67	50	150	2	-25	85	>1	TO-3, TO-5
LM320K-5.2	-6.2	-25	-5.2	50	10	67	50	150	2	0	70	>1	TO-3, TO-5
LM120K-12	-13	-30	-12	30	4	80	120	400	4	-55	125	>1	TO-3, TO-5
LM220K-12	-13	-30	-12	30	4	80	120	400	4	-25	85	>1	TO-3, TO-5
LM320K-12	-13	-30	-12	30	4	80	120	400	4	0	70	>1	TO-3, TO-5
LM120K-15	-16	-30	-15	30	5	80	150	400	4	-55	125	>1	TO-3, TO-5
LM220K-15	-16	-30	-15	30	5	80	150	400	4	-25	85	>1	TO-3, TO-5
LM320K-15	-16	-30	-15	30	5	80	150	400	4	0	70	>1	TO-3, TO-5

APPENDIX S

ELECTRONICS PACKAGES

Industry Package Cross-Reference Guide

	NSC	Signetics	Fairchild	Motorola	TI	RCA	Silicon General	AMD	Raytheon
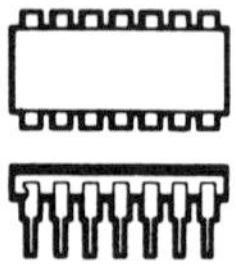 14/16 Lead Glass/Metal DIP	D	I	D	L		D	D	D	D, M
 Glass/Metal Flat-Pack	F	Q	F	F	F, S	K	F	F	J, F, Q
 TO-99, TO-100, TO-5	H	T, K, L, DB	H	G	L	S*, V1**	T	H	T, H
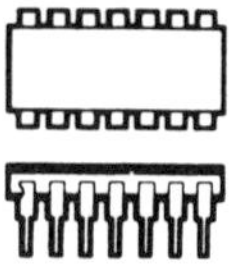 8, 14 and 16-Lead Low-Temperature Ceramic DIP	J	F	R, D	L	J				DC, DD
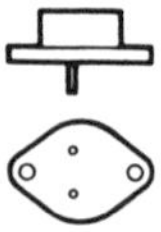 TO-3 (Steel)	K						K		K
TO-3 (Aluminum)	KC	DA	K	K	K		K		LK, TK
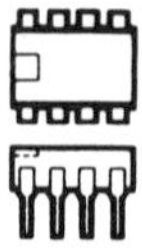 8, 14 and 16-Lead Plastic DIP	N	V, A, B	T, P	P	P, N	E	M, N	PC	N, DN, DP, MP

Figure S-1

		NSC	Signetics	Fairchild	Motorola	TI	RCA	Silicon General	AMD	Raytheon
(Package 37)	TO-202 (D-40, Durawatt)	P								
(Package 39)	"SGS" Type Power DIP	S		BP						
(Package 26)	TO-220	T	U	U		KC				
	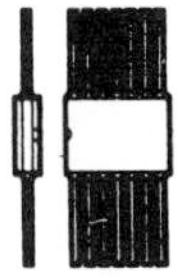 Low Temperature Glass Hermetic Flat Pack	W		F	F	W			FM	
(Package 38)	TO-92 (Plastic)	z	s	W	P	LP				

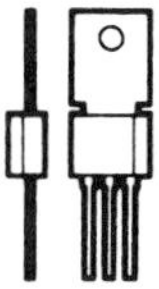

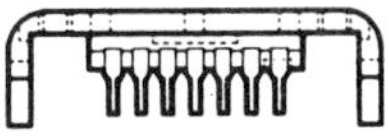

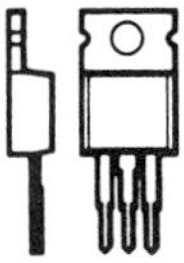

*With dual-in-line formed leads.
**With radially formed leads.

Courtesy of National Semiconductor Corporation

Figure S-2

APPENDIX T

POWER OP AMP SPECS

FEATURES

- WIDE SUPPLY RANGE — ±15V to ±150V
- HIGH OUTPUT CURRENT —
    1.5A Continuous (PB58)
    2.0A Continuous (PB58A)
- VOLTAGE AND CURRENT GAIN
- HIGH SLEW — 50V/μs Minimum (PB58)
    75V/μs Minimum (PB58A)
- PROGRAMMABLE OUTPUT CURRENT LIMIT
- HIGH POWER BANDWIDTH — 320 kHz Minimum
- LOW QUIESCENT CURRENT — 12mA Typical

APPLICATIONS

- HIGH VOLTAGE INSTRUMENTATION
- Electrostatic TRANSDUCERS & DEFLECTION
- Programmable Power Supplies Up to 280V p-p

DESCRIPTION

The PB58 is a high voltage, high current amplifier designed to provide voltage and current gain for a small signal, general purpose op amp. Including the power booster within the feedback loop of the driver amplifier results in a composite amplifier with the accuracy of the driver and the extended output voltage range and current capability of the booster. The PB58 can also be used without a driver in some applications, requiring only an external current limit resistor to function properly.

The output stage utilizes complementary MOSFETs, providing symmetrical output impedance and eliminating secondary breakdown limitations imposed by Bipolar Transistors. Internal feedback and gainset resistors are provided for a pin-strapable gain of 3. Additional gain can be achieved with a single external resistor. Compensation is not required for most driver/gain configurations, but can be accomplished with a single external capacitor. Enormous flexibility is provided through the choice of driver amplifier, current limit, supply voltage, voltage gain, and compensation.

This hybrid circuit utilizes a beryllia (BeO) substrate, thick film resistors, ceramic capacitors and semiconductor chips to maximize reliability, minimize size and give top performance. Ultrasonically bonded aluminum wires provide reliable interconnections at all operating temperatures. The 8-pin TO-3 package is electrically isolated and hermetically sealed using one-shot resistance welding. The use of compressible isolation washers may void the warranty.

TYPICAL APPLICATION

Figure 1. Inverting composite amplifier.

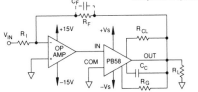

EQUIVALENT SCHEMATIC

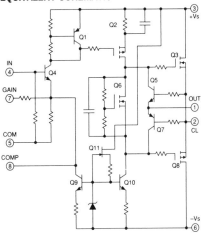

EXTERNAL CONNECTIONS

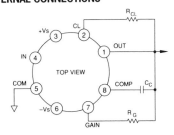

Figure T-1

ABSOLUTE MAXIMUM RATINGS

SUPPLY VOLTAGE, $+V_s$ to $-V_s$	300V
OUTPUT CURRENT, within SOA	2.0A
POWER DISSIPATION, internal at $T_C = 25°C$[1]	83W
INPUT VOLTAGE, referred to common	±15V
INPUT VOLTAGE, referred to $+V_s$	$+V_s$ −6.5V
TEMPERATURE, pin solder—10 sec max	300°C
TEMPERATURE, junction[1]	175°C
TEMPERATURE, storage	−65 to +150°C
OPERATING TEMPERATURE RANGE, case	−55 to +125°C

SPECIFICATIONS

PARAMETER	TEST CONDITIONS[2]	PB58 MIN	PB58 TYP	PB58 MAX	PB58A MIN	PB58A TYP	PB58A MAX	UNITS
INPUT								
OFFSET VOLTAGE, initial			±75	±1.75			±1.0	V
OFFSET VOLTAGE, vs. temperature	Full temperature range[3]		−4.5	−7		*	*	mV/°C
INPUT IMPEDANCE, DC		25	50		*			kΩ
INPUT CAPACITANCE			3			*		pF
INPUT VOLTAGE RANGE	Referred to common			±15			*	V
CLOSED LOOP GAIN RANGE		3	10	25	*		*	V/V
GAIN ACCURACY, internal Rg, Rf	$A_v = 3$		±10	±15		*	*	%
GAIN ACCURACY, external Rf	$A_v = 10$		±15	±25			*	%
PHASE SHIFT	f = 10kHz, $AV_{CL} = 10$, $C_C = 22pF$		10			*		°
	f = 200kHz, $AV_{CL} = 10$, $C_C = 22pF$		60			*		°
OUTPUT								
VOLTAGE SWING	Io = 1.5A (PB58), 2A (PB58A)	Vs−11	Vs −8		Vs−12	Vs−9		V
VOLTAGE SWING	Io = 1A	Vs−10	Vs −7		*	*		V
VOLTAGE SWING	Io = .1A	Vs−8	Vs −5		*	*		V
CURRENT, continuous		1.5			2.0			A
SLEW RATE	Full temperature range	50	100		75	*		V/µs
CAPACITIVE LOAD	Full temperature range		2200			*		pF
SETTLING TIME to .1%	$R_L = 100Ω$, 2V step		2			*		µs
POWER BANDWIDTH	$V_C = 100$ Vpp	160	320		240	*		kHz
SMALL SIGNAL BANDWIDTH	$C_C = 22pF$, $A_v = 25$, Vcc = ±100		100			*		kHz
SMALL SIGNAL BANDWIDTH	$C_C = 22pF$, $A_v = 3$, Vcc = ±30		1			*		MHz
POWER SUPPLY								
VOLTAGE, ±Vs[4]	Full temperature range	±15[6]	±60	±150	*	*	*	V
CURRENT, quiescent	Vs = ±15		11			*		mA
	Vs = ±60		12			*		mA
	Vs = ±150		14	18		*	*	mA
THERMAL								
RESISTANCE, AC junction to case[5]	Full temp. range, f > 60Hz		1.2	1.3		*	*	°C/W
RESISTANCE, DC junction to case	Full temp. range, f < 60Hz		1.6	1.8		*	*	°C/W
RESISTANCE, junction to air	Full temperature range		30			*		°C/W
TEMPERATURE RANGE, case	Meets full range specifications	−25	25	85	*	*	*	°C

NOTES:
* The specification of PB58A is identical to the specification for PB58 in applicable column to the left.
1. Long term operation at the maximum junction temperature will result in reduced product life. Derate internal power dissipation to achieve high MTTF (Mean Time to Failure).
2. The power supply voltage specified under typical (TYP) applies, $T_C = 25°C$ unless otherwise noted.
3. Guaranteed by design but not tested.
4. +Vs and −Vs denote the positive and negative supply rail respectively.
5. Rating applies if the output current alternates between both output transistors at a rate faster than 60Hz.
6. −V_s must be at least 30V below common.

CAUTION The PB58 is constructed from MOSFET transistors. ESD handling procedures must be observed.

The internal substrate contains beryllia (BeO). Do not break the seal. If accidentally broken, do not crush, machine, or subject to temperatures in excess of 850°C to avoid generating toxic fumes.

Figure T-2

APPENDIX U

DERIVATION OF EQS. (12-3) AND (12-4) AVERAGE POWER DISSIPATION BY A POWER AMPLIFIER WITH A SINE-WAVE OUTPUT

The following assumes:

1. The load R_L is resistive.
2. The output voltage is sinusoidal and swings symmetrically between the rails and has a peak value of V_p; see Fig. U-1.
3. There is no crossover distortion.
4. The quiescent power dissipation P_{DQ} is negligible.

Referring to Fig. U-1, transistor Q_1 conducts on positive alternations and Q_2 on the negative. In the time interval $0 \le t \le \pi/\omega$, Q_1 conducts, causing an instantaneous load voltage

$$v_L = V_p \sin \omega t. \qquad (U-1)$$

The load current is the collector current and can be expressed as

$$i_L = i_c = \frac{V_p \sin \omega t}{R_L}. \qquad (U-2)$$

Transistor Q_1's collector-to-emitter voltage v_{ce} is the difference in the rail voltage V_S and and the load voltage V_L. Thus

$$v_{ce} = V_S - V_p \sin \omega t. \qquad (U-3)$$

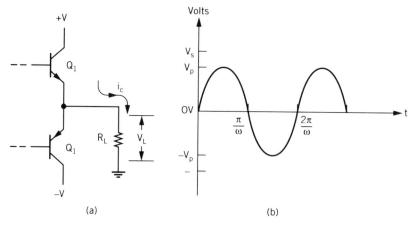

Figure U-1   ac Output of power amplifier; (a) complementary-pair BJTs, (b) symmetrical waveform; since $\omega = 2\pi f$, then $f = \omega/2\pi$ and $T = 2\pi/\omega$.

And, therefore, the instantaneous power dissipated by Q_1 is

$$p_D = v_{ce}i_c$$

or

$$p_D = (V_S - V_p \sin \omega t)\frac{V_p \sin \omega t}{R_L}. \tag{U-4}$$

By expanding, Eq. (U-4) becomes

$$p_D = \frac{1}{R_L}\left(V_S V_p \sin \omega t - V_p^2 \sin^2 \omega t\right). \tag{U-5}$$

The average power dissipation P_D is determined by integrating Eq. (U-5) over the interval that Q_1 conducts ($0 \le t \le \pi/\omega$) and then dividing by that time interval. Thus generally

$$P_D = \frac{1}{T}\int_0^\pi p_A \, dt,$$

and, more specifically,

$$P_D = \frac{\omega}{\pi}\int_0^{\pi/\omega} \frac{1}{R_L}\left(V_S V_p \sin \omega t - V_p^2 \sin^2 \omega t\right) dt. \tag{U-6}$$

From a table of integrals, we find that

$$\int \sin ax \, dx = \frac{-\cos ax}{a}$$

and

$$\int \sin^2 ax \, dx = \frac{x}{2} - \frac{\sin 2ax}{4a}.$$

Letting $x = t$ and then substituting these into Eq. (U-6) yields

$$P_D = \frac{\omega}{\pi R_L} \left[V_S V_p \left(\frac{-\cos \omega t}{\omega} \right) - V_p^2 \left(\frac{t}{2} - \frac{\sin^2 \omega t}{4\omega} \right) \right]_0^{\pi/\omega},$$

which evaluates to

$$P_D = \frac{1}{\pi R_L} \left[2V_S V_p - \pi V_p^2/2 \right]. \tag{U-7}$$

We find the maximum value on a P_D vs. V_p curve by differentiating the P_D equation and setting the derivative equal to zero. Thus

$$\frac{dP_D}{dV_p} = \frac{1}{\pi R_L} \left[2V_S - \pi V_p \right]. \tag{U-8}$$

Setting the right side equal to zero, we find that

$$2V_S = \pi V_p,$$

and therefore,

$$V_p = \frac{2V_S}{\pi} \tag{12-3}$$

at $P_{D(max)}$. Finally, substituting this into Eq. (U-7) yields

$$P_{D(max)} = \frac{1}{\pi R_L} \left[\frac{4V_S^2}{\pi} - \frac{\pi 4V_S^2}{2\pi^2} \right], \tag{U-7}$$

which simplifies to

$$P_{D(max)} = \frac{2V_S^2}{\pi^2 R_L}. \tag{12-4}$$

This equation is applicable to negative alternation too; that is, the interval $\pi/\omega \le t \le 2\pi/\omega$ in Fig. U-1, and therefore, it solves for the maximum power dissipation of the amplifier including both BJTs.

APPENDIX V

OP AMP APPLICATIONS

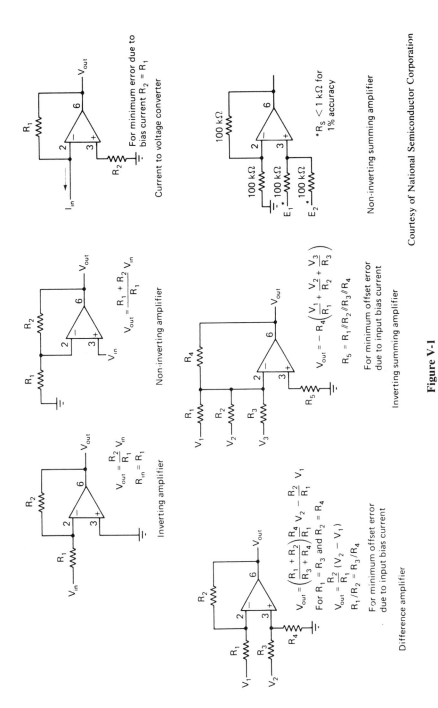

Current to voltage converter

For minimum error due to bias current $R_2 = R_1$

Non-inverting amplifier

$$V_{out} = \frac{R_1 + R_2}{R_1} V_{in}$$

Inverting amplifier

$$V_{out} = \frac{R_2}{R_1} V_{in}$$
$$R_{in} = R_1$$

Non-inverting summing amplifier

*$R_S < 1$ kΩ for 1% accuracy

Inverting summing amplifier

$$V_{out} = - R_4 \left(\frac{V_1}{R_1} + \frac{V_2}{R_2} + \frac{V_3}{R_3} \right)$$

$$R_5 = R_1 /\!/ R_2 /\!/ R_3 /\!/ R_4$$

For minimum offset error due to input bias current

Difference amplifier

$$V_{out} = \left(\frac{R_1 + R_2}{R_3 + R_4} \right) \frac{R_4}{R_1} V_2 - \frac{R_2}{R_1} V_1$$

For $R_1 = R_3$ and $R_2 = R_4$

$$V_{out} = \frac{R_2}{R_1} (V_2 - V_1)$$

$$R_1 / R_2 = R_3 / R_4$$

For minimum offset error due to input bias current

Figure V-1

Courtesy of National Semiconductor Corporation

329

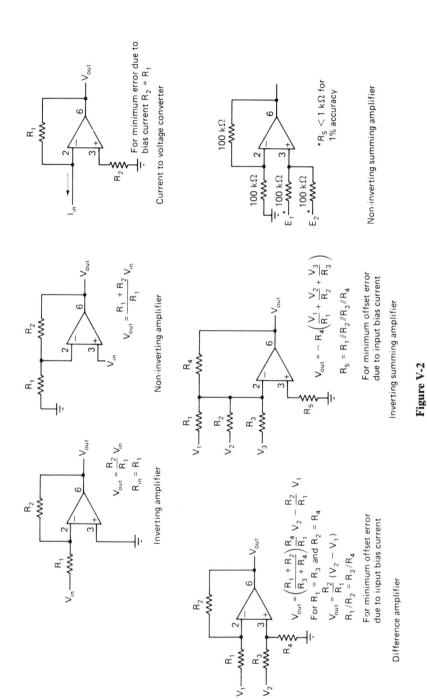

Current to voltage converter

For minimum error due to bias current $R_2 = R_1$

Non-inverting amplifier

$$V_{out} = \frac{R_1 + R_2}{R_1} V_{in}$$

Inverting amplifier

$$V_{out} = \frac{R_2}{R_1} V_{in}$$

$$R_{in} = R_1$$

Non-inverting summing amplifier

$^*R_S < 1\ k\Omega$ for 1% accuracy

Inverting summing amplifier

$$V_{out} = -R_4 \left(\frac{V_1}{R_1} + \frac{V_2}{R_2} + \frac{V_3}{R_3} \right)$$

$$R_5 = R_1 /\!/ R_2 /\!/ R_3 /\!/ R_4$$

For minimum offset error due to input bias current

Difference amplifier

$$V_{out} = \left(\frac{R_1 + R_2}{R_3 + R_4} \right) \frac{R_4}{R_1} V_2 - \frac{R_2}{R_1} V_1$$

For $R_1 = R_3$ and $R_2 = R_4$

$$V_{out} = \frac{R_2}{R_1} (V_2 - V_1)$$

$$R_1 / R_2 = R_3 / R_4$$

For minimum offset error due to input bias current

Figure V-2

330

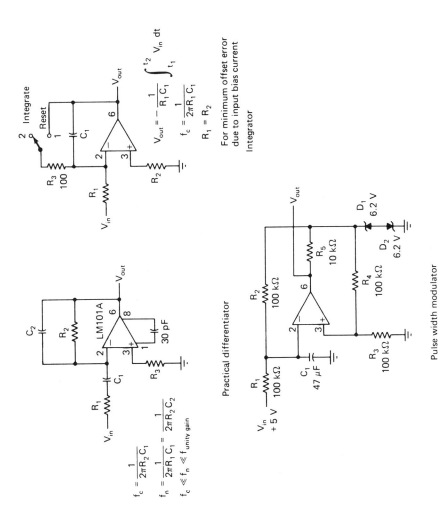

Integrator

$$V_{out} = -\frac{1}{R_1 C_1} \int_{t_1}^{t_2} V_{in}\, dt$$

$$f_c = \frac{1}{2\pi R_1 C_1}$$

$$R_1 = R_2$$

For minimum offset error
due to input bias current

Practical differentiator

$$f_c = \frac{1}{2\pi R_2 C_1}$$

$$f_n = \frac{1}{2\pi R_1 C_1} = \frac{1}{2\pi R_2 C_2}$$

$$f_c \ll f_n \ll f_{unity\ gain}$$

Pulse width modulator

Figure V-3

331

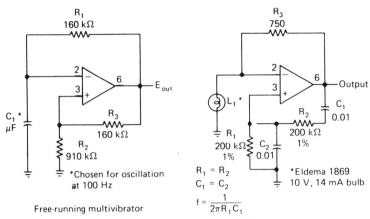

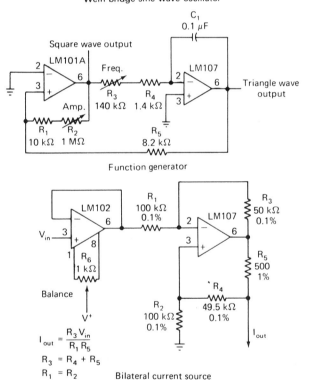

Figure V-4

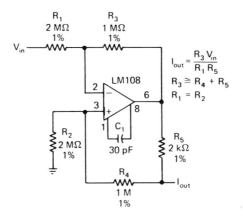

$$I_{out} = \frac{R_3 V_{in}}{R_1 R_5}$$

$$R_3 \cong R_4 + R_5$$

$$R_1 = R_2$$

Bilateral current source

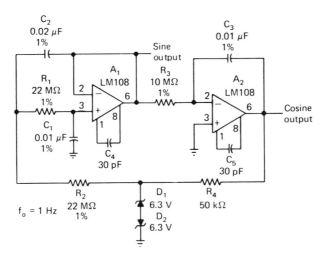

$f_o = 1$ Hz

Low frequency sine wave generator with quadrature output

Figure V-5

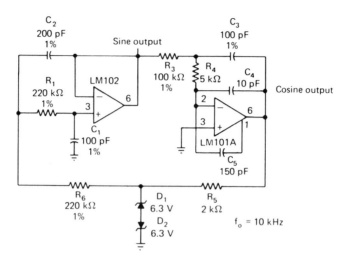

High frequency sine wave generator with quadrature output

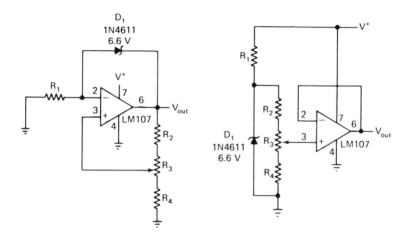

Positive voltage reference

Positive voltage reference

Figure V-6

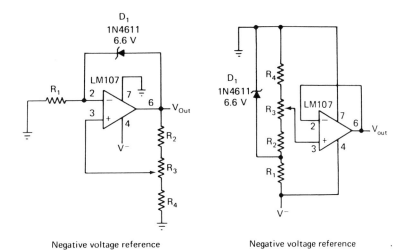

Negative voltage reference   Negative voltage reference

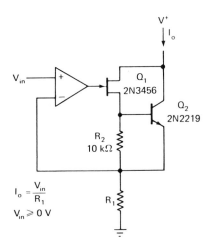

Precision current sink

Figure V-7

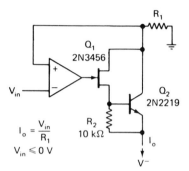

Precision current source

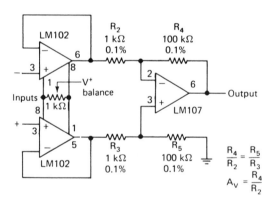

Differential-input instrumentation amplifier

Figure V-8

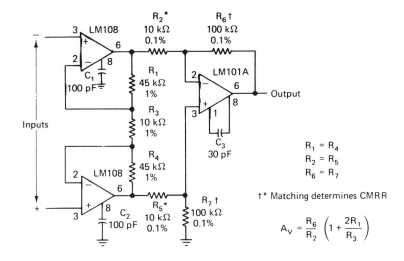

$R_1 = R_4$
$R_2 = R_5$
$R_6 = R_7$

†* Matching determines CMRR

$$A_V = \frac{R_6}{R_2}\left(1 + \frac{2R_1}{R_3}\right)$$

Differential input instrumentation amplifier
with high common mode rejection

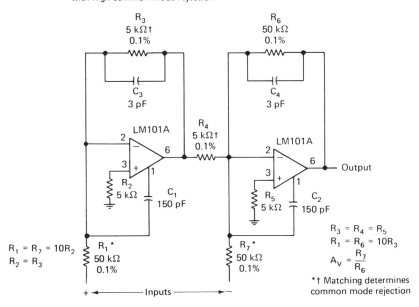

$R_3 = R_4 = R_5$
$R_1 = R_6 = 10R_3$
$$A_V = \frac{R_7}{R_6}$$

*† Matching determines
common mode rejection

Instrumentation amplifier with ± 100 volt common mode range

Figure V-9

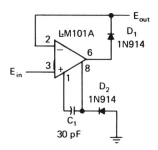

Precision diode

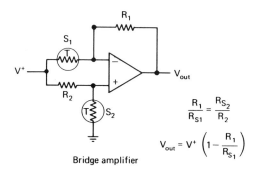

$$\frac{R_1}{R_{S1}} = \frac{R_{S_2}}{R_2}$$

$$V_{out} = V^+ \left(1 - \frac{R_1}{R_{S_1}}\right)$$

Bridge amplifier

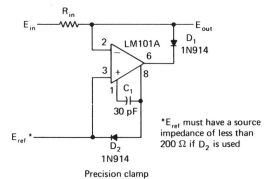

*E_{ref} must have a source
impedance of less than
200 Ω if D_2 is used

Precision clamp

Figure V-10

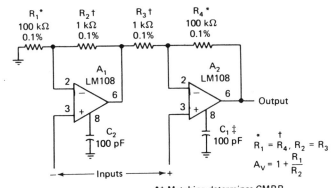

$$R_1 = R_4, R_2 = R_3$$

$$A_V = 1 + \frac{R_1}{R_2}$$

*† Matching determines CMRR
‡ May be deleted to maximize bandwidth

High input impedance instrumentation amplifier

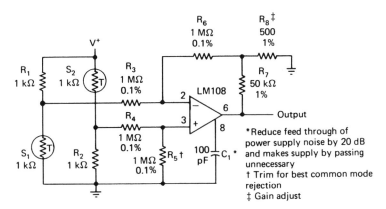

*Reduce feed through of
power supply noise by 20 dB
and makes supply by passing
unnecessary
† Trim for best common mode
rejection
‡ Gain adjust

Bridge amplifier with low noise compensation

Figure V-11

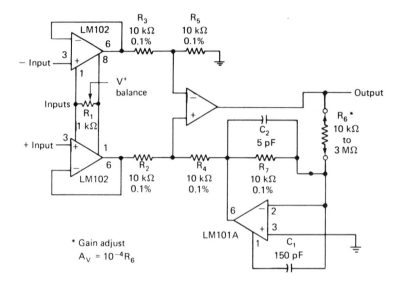

Variable gain, differential-input instrumentation amplifier

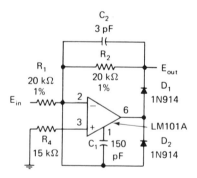

Fast half wave rectifier

Figure V-12

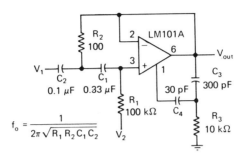

Tuned circuit

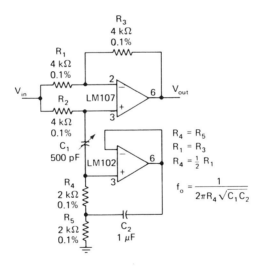

Easily tuned notch filter

Figure V-13

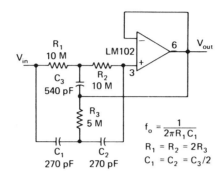

High Q notch filter

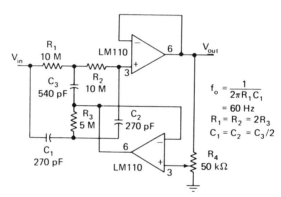

Adjustable Q notch filter

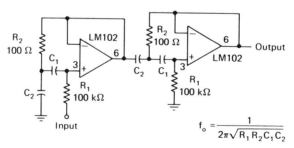

Two-stage tuned circuit

Figure V-14

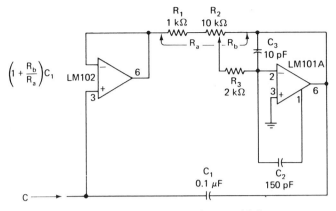

Variable capacitance multiplier

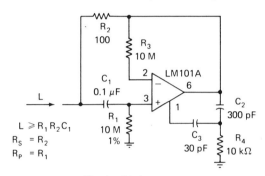

Simulated inductor

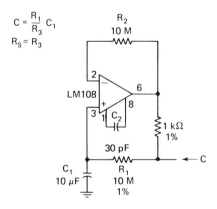

Capacitance multiplier

Figure V-15

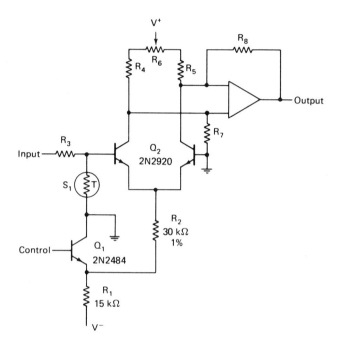

Voltage controlled gain circuit

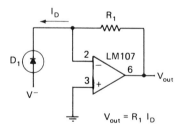

Photodiode amplifier

Figure V-16

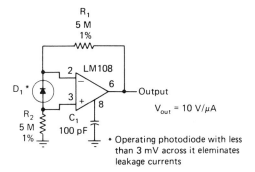

Photodiode amplifier

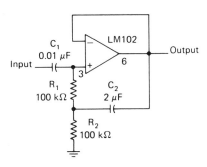

High input impedance ac follower

Figure V-17

ANSWERS TO SELECTED ODD-NUMBERED PROBLEMS

Chapter 1

1-1 $V_{id} = 0$ V, $V_{C_1} = V_{C_2} = 7.2$ V, $V_{od} = 0$ V.

1-3 $I_{C_1} \approx 1.795$ mA, $I_{C_2} \approx 2.2$ mA, $V_{C_2} \approx 6.4$ V, $V_{od} \approx 1.6$ V.

1-5 $V_{C_1} = V_{CC} = 15$ V.

1-7 $A_d = 260$, $A_V = 130$.

1-9 $V_{id} = 25$ mV, $V_{C_1} = 3.95$ V, $V_{C_2} = 10.45$ V, $V_{od} = 6.5$ V.

1-11 $I_o = 2.6$ mA.

1-13 $V_{D_1} = V_{D_2} = 5.48$ V, $V_{od} = 0$ V.

Chapter 2 (Review Questions)

2-1 Zero.

2-3 Infinite.

2-5 Zero.

2-9 Reduce, increase.

Chapter 3

3-1 (b).

3-3 (e).

3-5 $A_v = 2$, V_o has waveform (a).

3-7 $A_v = 101$, V_o has waveform (f).

3-9 As seen looking to the right of point x, the circuit of Fig. 3-17a has the more constant input resistance.

3-11 About 20 mV dc.

3-13 About 3 V dc.

3-15 $R_{o(\text{eff})} = 0.375\ \Omega$.

3-17 $V_o = 400$ mV dc.

3-19 $A_{v(\text{min})} = 2$, $A_{v(\text{max})} = 202$.

3-21 Waveform V_o is unclipped, 180° out of phase with the input but with peaks 150 times larger.

3-23 Waveform V_o is a sawtooth for the beginning alternations and is 180° out of phase with the input. It is clipped at plus or minus 5.5 V at times t_4, t_5, t_6, etc.

3-25 The output is in phase with the input and is not clipped on the first alternation that has a peak twice that of the input. The remaining alternations are clipped at ± 3 V.

3-27 The output is out of phase with the input and is not clipped on the first two alternations. The remaining alternations are clipped at about ± 13 V.

Chapter 4

4-1 -40.

4-3 -101.

4-5 Between 200 and 250 mV dc.

4-7 About 20 mV dc.

4-9 Between 220 and 280 mV dc.

4-11 990 Ω or about 1 kΩ.

4-13 About 20 mV dc.

4-15 Replace the 200-kΩ resistor with a smaller one or replace the 100-Ω resistor with a larger resistance value.

4-17 0.2 V.

4-19 0.3 V.

4-21   Provides capability to reduce (null) the output to 0 V.

4-23   About 6.2 V dc.

Chapter 5

5-1   80 dB.

5-3   (a) Output varies from 3 V to -2 V, (b) output noise is 1 V rms at 60 Hz because there is no CMRR with this circuit since it is not wired in differential mode.

5-5   (a) 1.82 V, (b) -5 V to 5 V, (c) 20-μV rms at 60 Hz.

5-7   (a) 9 V, (b) 8.18 V.

5-9   (a) 0 V dc, (b) -1.58 V to 1.43 V.

5-11   15 V to -15 V.

5-13   About 7.6/1.

5-15   11.

5-17 $V_{CM} = 5.3$ V, $V_o = 4.4$ V dc.

5-19   About 600 to 5700.

5-21   100.

5-23   (a) $A_v = A_{VOL}$, (b) $A_v = -4.4$.

5-25 R_a provides gain adjustment; R_4 provides NULL adjustment.

5-27   About 3.48 V.

Chapter 6

6-1   1 MHz.

6-3 $C_1 = 500$ pF, $R_1 = 1.5$ kΩ, $C_2 = 20$ pF.

6-5   1 MHz.

6-7   About 50 kHz.

6-9 $A_v = A_{VOL}$, BW is narrow.

6-11 -20 dB/decade or -6 dB/octave.

6-13   About 20 kHz.

6-15   0.2 μV rms at 60 Hz.

6-17 V_o is a sawtooth with 4-V peak-to-peak amplitude.

6-19 About 117 kHz.

6-21 $V_{no} \cong 55 \ \mu V$ rms.

6-23 10 kHz.

6-25 Almost square.

Chapter 7

7-1 $\Delta V_{io} \cong \pm 40 \ \mu V$, $\Delta V_{oo} \cong \pm 8$ mV.

7-3 About 8 μV rms.

7-5 10 $\mu V / V$.

7-7 $\Delta V_{oo} \cong 0.9$ mV.

7-9 The output drift from 500 mV is by about 0.25 mV in either case.

7-11 (a) About 399 mV, (b) about 400 mV.

7-13 (a) About 97.5 ns, (b) about 110 ns.

7-15 (a) About 136 mV for an LM741C or (b) 130 μV for an LM741E.

Chapter 8

8-1 $A_v = 21$.

8-3 $A_v = 11$, $R \cong 1.8$ kΩ.

8-5 $A_v = -1$.

8-7 (a) 9.4 V, (b) 5.55 V.

8-9 About 1 V.

8-11 About 22 V to 5.45 V.

8-13 $R_F / R_1 = 0.515$, $R_a / R_b < 1.73$, $R_s \cong 22$ Ω, $P_{C(max)} \cong 21$ W, max $P_z = 248$ mW.

8-15 -2.5 V, 540 Ω.

8-17 (a) 0.8 V, (b) 4 V.

Chapter 9

9-1 $f_c \cong 603$ Hz, $R_F = R = 1.2$ kΩ.

9-3 (a) -10 dB, (b) -5.1 dB, (c) -3.03 dB (d) -1.95 dB, (e) -1.35 dB.

9-5   0 dB down to 600 Hz. Below 600 Hz, roll-off is at a 20-dB/decade rate.

9-7 $d = 0.899$, 2 dB LP Chebyshev.

9-9 $f_x = f_p = 1505$ Hz.

9-11 $C_2 = 0.1 \ \mu F$, $R_1 = R_2 = 1406.74 \ \Omega$.

9-13 $d = 0.77$, 3-dB Chebyshev HP filter.

9-15 $f_p = 251.9$ Hz.

9-17 $R_1 = 7957.7 \ \Omega$, $R_2 = 418.8 \ \Omega$, $R_F = 159.155 \ k\Omega$, BW = 100 Hz, $f_1 = 951.28$ Hz, $f_2 = 1051.28$ Hz.

9-19   1031 Hz, 19.8 dB.

9-21 $R_1 = 39.78 \ k\Omega$, $R_2 = 812 \ \Omega$, $R_F = 79.58 \ k\Omega$, $R_3 = 1 \ k\Omega$, $R_4 = 1 \ k\Omega$, and $R_5 = 10 \ k\Omega$.

Chapter 10

10-1   Without the zeners, the output would be a sine wave with 10-V peaks. With these zeners, both the positive and negative alternations are clipped at about 6.6 V.

10-3   Without the diodes, the output would be sinusoidal, peaking at 11 V. With the diodes, positive alternations are clipped at about 4.6 V, but negative alternations are unclipped.

10-5   Zener D_1 is forward biased when the output attempts to swing negatively causing $V_o \cong 0.6$ V at such times. The positive alternations are sinusoidal and are clipped at about 4 V.

10-7   2000/1.

10-9   (a) $V_{o1} = -15$ V, $V_{o2} = 15$ V, $V_{o3} = 15$ V, (b) $V_{o1} = -15$ V, $V_{o2} = -15$ V, $V_{o3} = 15$ V, (c) $V_{o1} = V_{o2} = V_{o3} = -15$ V.

10-11   Waveform V_o is a square wave with the same frequency as the input sine wave. Waveform V_L is a differentiation of waveform V_o; i.e., V_L spikes positively when V_o rises positively; V_L spikes negatively when V_o changes negatively.

10-13   Like Fig. 10-10b, where $V_a' = -0.429$ V and $V_a = 0.429$. The noise is 0.857 V.

10-15   A square wave with no noise.

10-17   1 mA and 0 V.

10-19   The output be a full wave rectified waveform with a peak of 1.5 V.

10-21   (a) 1.2 V, (b) -2.8 V.

10-23 Output is rectangular with 4.74-ms positive alternations and 5.26-ms negative alternations.

Chapter 11

11-1 About 137 Hz.

11-3 26 V.

11-5 The sawtooth output V_o' will have its positive and negative alternations clipped.

11-7 V_o' will be a nonsymmetrical sawtooth with half of its alternations having a time period 20 times larger than the remaining alternations.

11-9 About 1.6 kΩ.

Chapter 12

12-1 $P_{D(\text{max})} \approx 13.8$ W.

12-3 $P_{D(\text{max})} \approx 17$ W.

12-5 $\theta_{SA} < 1.48°C$.

12-7 $R_{\text{LIM}} \approx 0.43$ Ω.

INDEX